3D Electro-Rotation of Single Cells

Synthesis Lectures on Biomedical Engineering

Editor

John D. Enderle, *University of Connecticut*

Lectures in Biomedical Engineering will be comprised of 75- to 150-page publications on advanced and state-of-the-art topics that span the field of biomedical engineering, from the atom and molecule to large diagnostic equipment. Each lecture covers, for that topic, the fundamental principles in a unified manner, develops underlying concepts needed for sequential material, and progresses to more advanced topics. Computer software and multimedia, when appropriate and available, are included for simulation, computation, visualization and design. The authors selected to write the lectures are leading experts on the subject who have extensive background in theory, application and design.

The series is designed to meet the demands of the 21st century technology and the rapid advancements in the all-encompassing field of biomedical engineering that includes biochemical processes, biomaterials, biomechanics, bioinstrumentation, physiological modeling, biosignal processing, bioinformatics, biocomplexity, medical and molecular imaging, rehabilitation engineering, biomimetic nano-electrokinetics, biosensors, biotechnology, clinical engineering, biomedical devices, drug discovery and delivery systems, tissue engineering, proteomics, functional genomics, and molecular and cellular engineering.

3D Electro-Rotation of Single Cells
Liang Huang and Wenhui Wang
November 2019

Spatiotemporal Modeling of Influenza: Partial Differential Equation Analysis in R
William E. Schiesser
May 2019

PDE Models for Atherosclerosis Computer Implementation in R
William E. Schiesser
November 2018

Computerized Analysis of Mammographic Images for Detection and Characterization of Breast Cancer

Paola Casti, Arianna Mencattini, Marcello Salmeri, and Rangaraj M. Rangayyan
June 2017

Models of Horizontal Eye Movements: Part 4, A Multiscale Neuron and Muscle Fiber-Based Linear Saccade Model
Alireza Ghahari and John D. Enderle
March 2015

Mechanical Testing for the Biomechanics Engineer: A Practical Guide
Marnie M. Saunders
January 2015

Models of Horizontal Eye Movements: Part 3, A Neuron and Muscle Based Linear Saccade Model
Alireza Ghahari and John D. Enderle
October 2014

Digital Image Processing for Ophthalmology: Detection and Modeling of Retinal Vascular Architecture
Faraz Oloumi, Rangaraj M. Rangayyan, and Anna L. Ells
January 2014

Biomedical Signals and Systems
Joseph V. Tranquillo
December 2013

Health Care Engineering, Part II: Research and Development in the Health Care Environment
Monique Frize
October 2013

Health Care Engineering, Part I: Clinical Engineering and Technology Management
Monique Frize
October 2013

Computer-aided Detection of Architectural Distortion in Prior Mammograms of Interval Cancer
Shantanu Banik, Rangaraj M. Rangayyan, and J.E. Leo Desautels
January 2013

Content-based Retrieval of Medical Images: Landmarking, Indexing, and Relevance Feedback

Paulo Mazzoncini de Azevedo-Marques, and Rangaraj Mandayam Rangayyan
January 2013

Chronobioengineering: Introduction to Biological Rhythms with Applications, Volume 1
Donald McEachron
October 2012

Medical Equipment Maintenance: Management and Oversight
Binseng Wang
October 2012

Fractal Analysis of Breast Masses in Mammograms
Thanh M. Cabral and Rangaraj M. Rangayyan
October 2012

Capstone Design Courses, Part II: Preparing Biomedical Engineers for the Real World
Jay R. Goldberg
September 2012

Ethics for Bioengineers
Monique Frize
November 2011

Computational Genomic Signatures
Ozkan Ufuk Nalbantoglu and Khalid Sayood
May 2011

Digital Image Processing for Ophthalmology: Detection of the Optic Nerve Head
Xiaolu Zhu, Rangaraj M. Rangayyan, and Anna L. Ells
January 2011

Modeling and Analysis of Shape with Applications in Computer-Aided Diagnosis of Breast Cancer
Denise Guliato and Rangaraj M. Rangayyan
January 2011

Analysis of Oriented Texture with Applications to the Detection of Architectural Distortion in Mammograms
Fábio J. Ayres, Rangaraj M. Rangayyan, and J. E. Leo Desautels
2010

3D Electro-Rotation of Single Cells
Liang Huang and Wenhui Wang

ISBN: 978-3-031-00538-1 Paperback
ISBN: 978-3-031-01666-0 eBook
ISBN: 978-3-031-00045-4 Hardcover

DOI 10.1007/978-3-031-01666-0

A Publication in the Springer Nature series
SYNTHESIS LECTURES ON ADVANCES IN AUTOMOTIVE TECHNOLOGY

Lecture #58
Series Editor: John D. Enderle, University of Connecticut

Series ISSN   1930-0328   Print   1930-0336   Electronic

3D Electro-Rotation of Single Cells

Liang Huang, Wenhui Wang*
State Key Laboratory of Precision Measurement Technology and Instrument
Department of Precision Instrument, Tsinghua University, Beijing, China

*Corresponding author: wwh@tsinghua.edu.cn

SYNTHESIS LECTURES ON BIOMEDICAL ENGINEERING #58

ABSTRACT

Dielectrophoresis microfluidic chips have been widely used in various biological applications due to their advantages of convenient operation, high throughput, and low cost. However, most of the DEP microfluidic chips are based on 2D planar electrodes which have some limitations, such as electric field attenuation, small effective working regions, and weak DEP forces. In order to overcome the limitations of 2D planar electrodes, two kinds of thick-electrode DEP chips were designed to realize manipulation and multi-parameter measurement of single cells.

Based on the multi-electrode structure of thick-electrode DEP, a single-cell 3D electro-rotation chip of "Armillary Sphere" was designed. The chip uses four thick electrodes and a bottom planar electrode to form an electric field chamber, which can control 3D rotation of single cells under different electric signal configurations. Electrical property measurement and 3D image reconstruction of single cells are achieved based on single-cell 3D rotation. This work overcomes the limitations of 2D planar electrodes and effectively solves the problem of unstable spatial position of single-cell samples, and provides a new platform for single-cell analysis.

Based on multi-electrode structure of thick-electrode DEP, a microfluidic chip with opto-electronic integration was presented. A dual-fiber optical stretcher embedded in thick electrodes can trap and stretch a single cell while the thick electrodes are used for single-cell rotation. Stretching and rotation manipulation gives the chip the ability to simultaneously measure mechanical and electrical properties of single cells, providing a versatile platform for single-cell analysis, further extending the application of thick-electrode DEP in biological manipulation and analysis.

KEYWORDS

thick electrodes, DEP, 3D rotation, optical stretcher, multi-parameter measurement

Contents

Acknowledgments

This work was supported by the NSFC (no. 61774095, 21727813), the National Key R&D Program (no. 2016YFC0900200), and the One-Thousand Young Talent Program of China.

CHAPTER 1

Introduction

1.1 OVERVIEW OF MICROFLUIDICS

1.1.1 BACKGROUND AND BRIEF DEVELOPMENT HISTORY

Microfluidics refers to the science and technology of manipulating fluids in microchannels [1]. Microfluidics can be used to integrate basic operation units such as sample preparation, reaction, separation, and detection in biological, chemical, and medical analysis processes onto a micrometer-scale chip to form miniaturized total analytical systems (μTAS) or *Lab on a Chip* to automate the analysis process. Microfluidics has the advantages of low cost, small volume, low sample and reagent requirements, and fast reaction. It has been widely used in biomedicine, food safety, environmental monitoring, and other application fields. Figure 1.1(a) shows a portable, disc-based, and fully automated enzyme-linked immuno-sorbent assay (ELISA) system [2]. Figure 1.1(b) shows the microfluidic chip for single circulating tumor cells (CTC) isolation and whole genomic amplification [3].

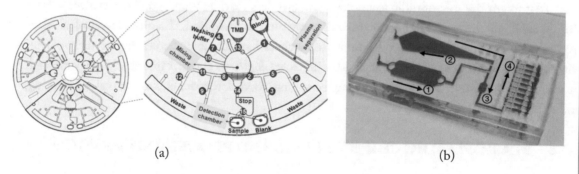

(a)   (b)

Figure 1.1: Examples of microfluidic chips: (a) a portable, disc-based, and fully automated enzyme-linked immuno-sorbent assay (ELISA) system [2], used with permission from the Royal Society of Chemistry; and (b) a microfluidic chip for single CTC isolation and whole genomic amplification [3], used with permission from the Royal Society of Chemistry.

The concept of microfluidic analysis systems was first proposed by Manz and Widmer in 1990, emphasizing the "micro" and "total" aspects of the analysis system and the chip fabrication methods based on MEMS technologies [4], and in 1992 the capillary electrophoresis experiment was successfully implemented on the microchip [5].

Subsequently, Harrison et al. implemented a capillary electrophoresis injection analysis experiment using a micro-scale chip in 1993, and set the main configuration of the microfluidic system to a chip with a thickness of generally no more than 5 mm and an area of several square millimeters to several square centimeters [6]. Subsequently, microfluidic technologies have developed rapidly. Woolley and colleagues first realized the DNA sequencing using microfluidic technology in 1995 [7], and realized the on-line detection of PCR products using microfluidic technology during 1996–1998 [8] and chemiluminescence detection experiments [9]. In 1998, Bums reported on the multi-function microfluidic chip for DNA analysis, which integrates injection, mixing, reaction, electrophoretic separation, and optical detection [10], which opens up the versatility road of microfluidic chip. In 2002, Quake et al. published an article in *Science* which introduced microfluidic large-scale integrated design methods and applications [1]. In the following year, Weigl et al. reviewed the microfluidic operation methods such as microfluidic driving and cell manipulation [11]. In the same year, Erickson et al. conducted numerical simulations and experimental studies on microfluidic motion. The transmission properties of microfluidics are summarized and provided a theoretical basis for microfluidic chip design [12]. In 2006, Whiteside et al. published an article in *Nature* to review the development of microfluidic chips and point out the direction and trend of microfluidic development in the future [13]. In 2014, Bhatia et al. published the work of organ chips in *Nature Biotechnology* [14] which was used to simulate the microenvironment and function of human organs, and further promoted the application development of microfluidic technologies.

Although the microfluidic technology emerged less than 30 years ago, it has evolved from a simple capillary electrophoresis miniaturization method to biological, chemical, mechanical, electronic, material, and medical applications. Researchers have combined microfluidic technologies with optical, mechanical, electrical, and acoustic technologies to conduct interdisciplinary research in multiple application fields, such as drug screening, food detection, environmental monitoring, and aerospace science.

1.1.2   MICROFLUIDIC CHIP MATERIAL AND PROCESSING METHOD

Microfluidic chip material

The composition of the microfluidic chip system mainly includes microchannels, microactuators, microcontrollers, micro-sensors components, and connectors [15]. Material properties play key roles in microfluidic systems, such as the hydrophobicity, light transmission, conductivity, and biocompatibility of the microchannel surface. At present, the materials commonly used in microfluidic chips are mainly silicon materials, glass materials, and high molecular polymer materials.

(a) Silicon and glass materials

Silicon materials such as silicon wafers were the first materials to be used in microfluidic chips [16]. However, silicon materials have limitations in their application. First, the light transmittance is poor, which limits the observation and detection of the sample, and secondly, the electrical insulating properties are poor, and the cost of the silicon material is high. Later, researchers began to use glass materials in microfluidic chips to replace pure silicon materials [17, 18].

Compared to silicon materials, glass materials have advantages of high light transmittance and high biocompatibility. Common glass types include chrome glass, boride glass, indium tin oxide (ITO), quartz glass, etc., which can be selected according to application requirements. For example, ITO glass with the characteristics of transparent electrode, high electrical conductivity, and high light transmission has been widely used to fabricate microelectrodes in microfluidic chips [19]. Quartz glass has a weak absorption of ultraviolet light, and is suitable for applications requiring UV absorption detection [20].

(b) High molecular polymer materials

In addition to silicon materials and glass materials, polymer materials are widely used in microfluidic chips due to their high light transmittance, electrical insulation, chemical inertness, and low cost [21]. The polymer materials commonly used in microfluidic chips are thermoplastic polymers and curable polymers.

Thermoplastic polymers mainly include polyamide (PI) [22], polycarbonate (PC) [23, 24], and polymethyl methacrylate (PMMA) [25, 26]. Such materials can be thermoformed by using a mold, which has the characteristics of high light transmittance, low cost, and long service life, but the surface hardness is poor, and scratches are easily affected to observe the experiment.

Curable polymers mainly include polydimethylsiloxane (PDMS), epoxy resin, and polyurethane. Due to the high light transmittance, high biocompatibility, and convenient processing, PDMS has become one of the most widely used high polymer materials in the field of microfluidic chips [27–29].

(c) Other materials

In addition to the above two types of materials, in recent years paper materials [30–32] and printed circuit boards (PCB) [33, 34] have gradually been applied in microfluidics in recent years. The paper materials mainly use the capillary effect to realize micro-liquid driving, and the surface hydrophilic/hydrophobic modification can realize the function of the microchannel on the paper, and are mostly used in real-time detection [35]. PCB materials with the advantages of PCB microelectrode processing are mostly used as electrodes for electrical control and detection of the sample [36].

In short, the materials of the microfluidic chip are the premise and basis of the microfluidic technology. In practical applications, the chip materials need to be selected according to the application requirements. The factors to be considered are: (a) the difficulty of processing the material, the processing accuracy and cost; (b) physical properties of the material, such as insulation, transparency, and thermal conductivity; and (c) chemical properties of the material, such as surface hydrophobicity and biocompatibility.

Processing technology of microfluidic chips

MEMS technologies are the most common methods of microfluidic chip processing. The processing technologies for fabricating a 2D planar structure includes photolithography [37], oxidation [38], chemical vapor deposition (CVD) growth [39], coating [40], etc. The processing technology for fabricating the 3D structure mainly includes photolithography, chemical etching, plasma etching [41, 42], X-ray lithography (Lithography, Electroplating, and Molding, LIGA) [43, 44], and so on.

(a) Processing technologies of silicon and glass materials

Silicon materials are mainly used for processing liquid flow drive and control devices such as micropumps and microvalves, or as a positive mold for processing high molecular polymer materials in hot pressing and molding. It is often processed by photolithography and etching techniques. It usually consists of a film deposition, photolithography, and etching process. In the processing of quartz and glass, the surface is often modified by chemical methods, and then microchannel processing is performed using photolithography and etching techniques.

(b) High polymer materials processing technology

The processing methods of high molecular polymer materials mainly include: molding methods [45], hot pressing methods [46], LIGA technologies [47], and soft lithographies [48, 49]. The molding methods first form a convex mold of the channel by photolithography and etching, and then pour polymer material on the mold, and finally cure the high molecular polymer material, and peel off to obtain a chip having microchannels. The hot pressing methods bond the high molecular polymer material and the mold together in the hot pressing device. When the high molecular polymer is softened by heating, the corresponding microstructure can be printed on the mold, and demolding is performed to obtain a chip having microchannels after cooling.

LIGA technologies are suitable for fabricating high aspect ratio polymer chip structures, including X-ray deep lithographies, electroforming, and injection molding. X-ray deep lithographies can obtain a high aspect ratio microchannel structure in the photoresist, electroforming mold deposits metal in the gap of the developed photoresist image, and lift-off the photoresist to obtain the high aspect ratio structure. Injection molding is the formation of the microchannel structure on the polymer material by replica molding on a mold.

Soft lithography is a new method of the micro-pattern replication process. The core is to fabricate micro-structured chips through lithography and molding processes. It has the advantages of low-cost and convenient processing, and is widely used in biotechnology, sensors, biotechnology, and other fields. The processing steps of the mold are mainly as follows: first, a layer of photoresist is spin-coated on the hard substrates, then pre-baked to volatilize and solidify the solvent in the photoresist, then subject to mask exposure and post-baking treatment, and finally a mold with microstructure is developed.

With more chip materials discovered, the processing technologies of microfluidic chips are constantly updated. As the mainstream materials for processing microfluidic chips, the molding process of polymer materials will be further studied, and its application range will be broader.

1.2   SAMPLE MANIPULATION METHODS IN MICROFLUIDIC CHIPS

The microfluidic chip controls the flow of microfluids through microfluidic devices such as micro-channels, microarrays, and microchambers; often combined with fluidic methods [50, 51], magnetic methods [52, 53], acoustic methods [54, 55], optical methods [56, 57], electrical methods [58, 59], and other technical means to achieve micro-operation of biological samples.

1.2.1   FLUIDIC METHODS

Fluidic methods utilize special microstructures or microvalves in microfluidic channels to control microfluidic flow to achieve manipulation of cells, such as single-cell capture [60, 61] and cell separation [62, 63]. For example, Zhang et al. designed a hook-type single-cell capture structure in the microchannel, which uses a single-cell array capture structure to achieve capture and co-culture of two cells (Figure 1.2(a)) [64]. In order to improve the efficiency of single-cell capture, Jin et al. designed a ladder-type single-cell capture chip based on the principle of minimum flow resistance, with a single-cell capture efficiency of 86% (Figure 1.2(b)) [65]. Also based on the principle of minimum flow resistance, Mi et al. proposed a single-cell matrix capture structure (Figure 1.2(c)). The device enables efficient deterministic single-cell capture and flexible-cell capture patterning by designing array patterns of trap site [66].

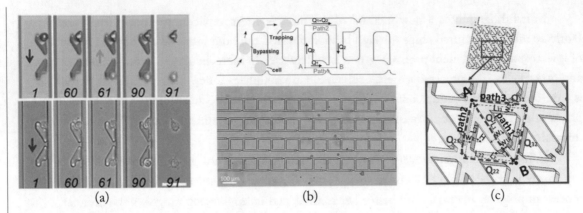

Figure 1.2: Single-cell capture structure: (a) hook-type capture structure [64]; (b) ladder-type capture structure [65]; and (c) single-cell matrix capture structure [66]. Used with permission from the Royal Society of Chemistry.

In cell separation, main methods include (a) inertial force [67, 68] and (b) deterministic lateral displacement (DLD) [69, 70]. The inertial force methods use the characteristics of different size cells in the spiral microchannel subject to different inertial forces and positional displacement to achieve cell separation. Since the technology is based on a continuous flow, it is possible to process a large number of samples in a short time. For example, Al-Halhouli et al. reported a label-free spiral flow chip that focusing efficiency of 99.1% of yeast cells (Figure 1.3(a)) [71]. The DLD methods are to design micropillars array structures with a certain displacement angle in the microchannel, and the continuous trajectory in the microchannel is different according to the difference size to achieve continuous separation. As cells flow along the streamline through the micropillars, larger cells will shift laterally, and smaller cells will continue to flow along the original streamline. Compared with the traditional membrane filtration methods, the DLD methods have the advantages of high throughput, label-free separation, and no clogging. For example, Liu et al. proposed a microfluidic chip (Figure 1.3(b)) [72] that combines DLD array and surface antibody modification technology to achieve rapid enrichment and capture of CTCs.

Although fluid methods can realize label-free and high throughput biological manipulation, the methods rely on the shape and structure of the microchannel, and depends on the differences of cell size that result in low specificity for cell sorting.

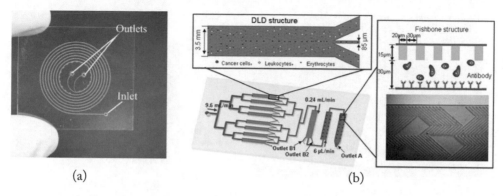

(a)   (b)

Figure 1.3: (a) Inertial force cell focusing chip [71]; and (b) DLD cell sorting chip [72].

1.2.2   OPTICAL METHODS

Optical tweezers utilize an optical gradient trap formed by a single beam to achieve capture and movement of single cells, which forms a 3D optical potential well when the laser beam is focused on the cell, thereby inducing optical pressure to capture the cell. By controlling the trajectory of the optical tweezers, it is possible to accurately trap and move cells [73, 74] or perform cell alignment patterning [75, 76]. Liberale et al. proposed a fiber-based optical tweezer structure that achieves stable single-cell 3D capture by micro-prism beam deflectors (Figure 1.4(a)) [77]. In general, only one single cell can be operated at a time. In order to achieve multi-cell simultaneous operation, Werner et al. proposed a microfluidic chip with refractive multiple optical tweezers (Figure 1.4(b)). This method was used to immobilize more than 200 yeast cells and perform patterning at a time [78].

Optical tweezers are also commonly used in cell sorting. According to the control method, they can be divided into passive and active sorting methods. The passive sorting method uses a fixed-point optical tweezers to achieve cell sorting in the microfluidic channel. Cells with different refractive indices, sizes, or shapes will be subject to positional shifts in the light field by different sizes of optical forces (Figure 1.4(c)) [79]. The passive sorting methods rely on the differences in the intrinsic properties of the cells, the sensitivity to cell selection and the sorting efficiency is low. Active sorting methods are based on differences in cell intrinsic properties and differences in biological properties, such as cell sizes and fluorescent label, using dynamic optical tweezers to classify cells. Although the accuracy of active sorting is higher than that of passive sorting, it needs to be combined with sensors to form a feedback control system (Figure 1.4(d)) [80].

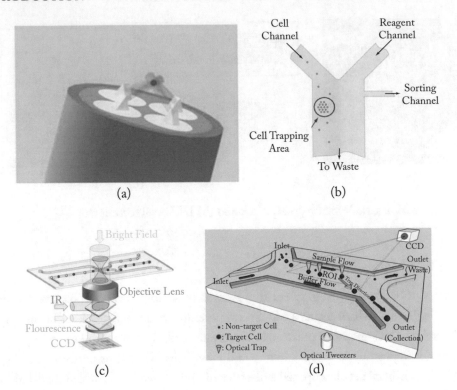

Figure 1.4: Optical trap methods: (a) fiber tweezers for 3D capture of single cells [77]; (b) array-type capture of single cells [78], based on and used with permission from the Royal Society of Chemistry; (c) passive sorting of microfluidic chips [79]; and (d) active sorting microfluidic chip [80]. Image (c) and (d) based on and used with permission from the Royal Society of Chemistry.

Although optical tweezers can accurately manipulate single cells, the laser beam is easy to damage cells, and require expensive optical system.

1.2.3   MAGNETIC METHODS

Generally, cells have very weak magnetic properties. Selectively modifying the surface of a cell to change its magnetic properties such that the cell can be micro-operated by an external magnetic field [81, 82]. The typical magnetic activated cell sorting methods can be used to separate cells. It is based on the combination of cell surface antigens and specific antibodies linked to magnetic beads. The external magnetic field can hold the cells modified with magnetic beads in the magnetic field, thereby realizing cell separation [83, 84]. Magnetic activated cell sorting methods are low cost, easy to operate, and highly sensitive. Huang et al. developed a ferromagnetic microfluidic system for enhanced CTC detection (Figure 1.5) [85]. Because the ferromagnetic micro-magnet is magnetized, the magnetic field will be enhanced, which can amplify the magnetic field strength in the micro-channel, strengthen the interaction between the modified CTC and the magnetic field, and successfully achieve the capture of four cancer cell lines, with capture efficiency >97%.

Magnetic activated cell sorting methods can efficiently manipulate the target cells, but the sample needs to be modified, and the peripheral magnetic field generating device is required for the control, which is inconvenient for miniaturization and integration of the chip.

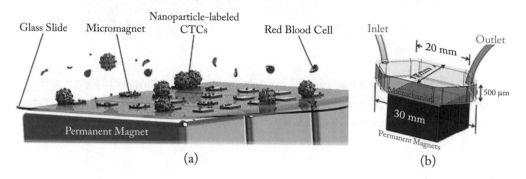

Figure 1.5: Ferromagnetic microfluidic system for enrichment of CTC [85]; (a) schematic diagram of chip operation; and (b) dimensional drawing of chip specific structure.

1.2.4 ACOUSTIC METHODS

Due to the advantages of non-contact operation, low-cost, high-controllability and high-biocompatibility, surface acoustic wave (SAW) has been widely used in cell separation [86, 87] and cell localization [88]. Acoustic methods generally realize the micro-operation of cells by forming an acoustic field in the microchannel by the piezoelectric transducer. Under the action of the acoustic radiation force the cells move to the position of the node (the position with the smallest amplitude) or the anti-node (the position with the largest amplitude). Cells with different physical properties such as sizes and densities are subject to acoustic radiation forces of different magnitudes and directions. The acoustic field distribution can be adjusted and controlled by changing the shape and position of the piezoelectric transducer. For example, Li et al. proposed a microfluidic chip based on SAW method, which can separate high concentration of CTC cells (~100 cells/ml) from peripheral blood (Figure 1.6(a)) [89]. In order to achieve single-cell sorting, Collins et al. proposed a focused SAW method that uses a focused interdigital transducer (FIDTs) to generate a local acoustic field for single-cell manipulation (Figure 1.6(b)) [90].

Acoustic methods can control cells without labels and damage, but it depends on the shape of the microchannel and difficult to separate cells of similar size and density.

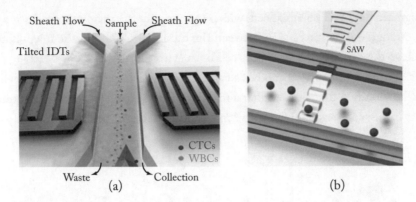

Figure 1.6: Surface acoustic wave microfluidic chip for cell sorting: (a) symmetrical forked surface acoustic wave cell sorting chip [89]; and (b) fFocused surface acoustic wave single-cell sorting chip [90]. Used with permission from the Royal Society of Chemistry.

1.2.5 DEP METHODS

Because of its label-free, low damage, high efficiency and easy operation, dielectrophoresis (DEP) technologies have been widely used in cell capture [91, 92], cell fusion [93, 94], and cell sorting [95, 96]. Due to the dielectric properties of cells, they could be polarized under non-uniform electric fields and formed electric dipoles, which are subjected to DEP forces or torques. In association with the differences in electrical properties of cells, DEP force or torque differs across cells under the same electric field. DEP methods can control the strength and direction of DEP force and torque by changing electrical signal parameters (amplitude, frequency and phase). Song et al. proposed a microfluidic chip for sorting stem cells (Figure 1.7(a)) [97]. There is an angle between the DEP force generated by the parallel electrodes and the flow force. Under the action of DEP force and flow force, different types of cells will have different offset distances at the exit, which can realize sorting of different cells. The chip has a collection efficiency of 92% for human mesenchymal stem cells along with purity up to 84%.

As opposed to traditional DEP methods, insulating pillar structures in microchannel were also designed to produce DEP effect called insulator-based DEP (iDEP) [98, 99]. The insulating pillar structures change the electric field distribution in the microchannel and can generate DEP force at the edge of the insulating pillars. LaLonde et al. proposed a cell enrichment microfluidic chip based on iDEP, as shown in Figure 1.7(b) [100]. The chip successfully captured and enriched yeast cells with capture efficiency > 99%.

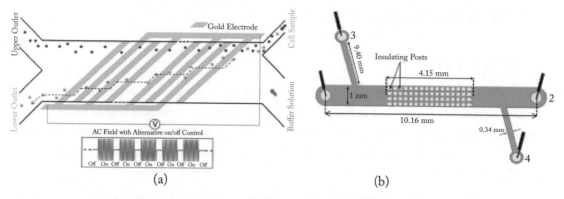

Figure 1.7: DEP microfluidic chip for cell sorting: (a) parallel electrode DEP cell sorting chip [97] (used with permission from the Royal Society of Chemistry; and (b) insulating DEP cell enrichment chip [100] (used with permission from AIP Publishing).

The strength and direction of DEP force and torque are related to the cell, solution, and electrical signal parameters. By adjusting electrical signal parameters, precise manipulation of single cells and cell population can be achieved. As MEMS technologies advance, the electrodes in microfluidic chips can be made more sophisticated, extending from 2D planar electrodes to thick-electrode structures. And the microfluidic chip based on the DEP technology can be easily combined with other various technologies to realize multifunctional operations.

The above sample manipulation mechanisms are summarized in Table 1.1.

Table 1.1: Biological control mechanisms in microfluidic chips			
Methods	**Theory**	**Throughput**	**Application**
Fluidic	Microstructures or microvalves in microchannels are used to control microfluidics to control the biological control	High	Cell capture, separation
Optical	The optical gradient well formed by a single beam is used to capture and move single cells	Low	Cell capture, Translation, Sorting
Magnetic	Cell surface antigens are combined with specific antibodies attached to magnetic beads to manipulate cells using an external magnetic field	High	Cell enrichment
Acoustic	The micro-operation of cells is realized by the acoustic field formed in the microchannel by piezoelectric transducer device	High	Cell separation, cell trap
DEP	The polarization of cells are formed electric dipole under non-uniform electric field, which is subjected to DEP force or torque	High	Sorting, Separation, Electro-rotation

1.3   DEP MICROFLUIDIC CHIPS

1.3.1   THEORY OF DEP

When a particle is polarized in an electric field, the charges in the particle are redistributed. The positive and negative charges move in opposite directions, causing a heterogeneous induced charge at the interface between the particle and the solution [101]. Due to the different polarizabilities of the solution and particle, the amount of induced charges generated at their interface is different, as shown in Figure 1.8.

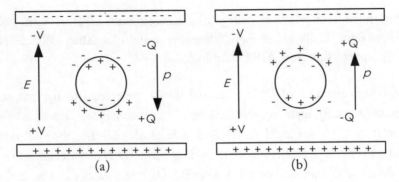

Figure 1.8: Schematic diagram of particle polarization: (a) the particle polarizability is lower than the solution; and (b) the particle polarizability is larger than the solution.

If the polarizability of the solution is greater than that of the particle, the amount of induced charges generated by the solution polarization is more than that of the particle, so the lower interface of the particle will be positively charged and the upper interface will be negatively charged, as shown in Figure 1.8(a). On the contrary, the lower side interface of the particle is negatively charged and the upper side is positively charged, as shown in Figure 1.8(b).

A charged system consisting of two equal-point charge +q and -q at a distance d is called an electric dipole, as shown in Figure 1.9. Under the action of an electric field, the particle is polarized to form an electric dipole. When the distance d between the set point charges +q and -q is much smaller than u (the distance of the negative charge from the origin point), the dipole moment $\vec{p}$ (C·m) can be expressed as:

$$\vec{p} = q\vec{d} \cdot$$

(1-1)

The direction is a negative charge pointing to a positive charge.

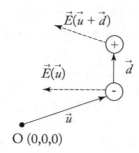

Figure 1.9: Schematic diagram of dipole moment.

Under the action of the non-uniform electric field E, the resultant force F of the electric field of the electric dipole is

$$\vec{F} = q\vec{E}(\vec{u}+\vec{d}) - q\vec{E}(\vec{u}). \qquad (1\text{-}2)$$

The first term on the right-hand side of Equation (1-2) is subjected to Taylor series expansion and after omitting the high order term one can get

$$q\vec{E}(\vec{u}+\vec{d}) = q\vec{E}(\vec{u}) + q\vec{d} \cdot \nabla\vec{E}(\vec{u}). \qquad (1\text{-}3)$$

Combining Equations (1-2) and (1-3) yields

$$\vec{F} = q\vec{d} \cdot \nabla\vec{E}. \qquad (1\text{-}4)$$

Substituting Equation (1-1) into (1-4)

$$\vec{F} = \vec{p} \cdot \nabla\vec{E}. \qquad (1\text{-}5)$$

For a particle with radius r_p, the dipole moment can be expressed as [102]

$$\vec{p} = 4\pi r_p^3 \, \varepsilon_m \cdot K_{CM} \cdot \nabla\vec{E}, \qquad (1\text{-}6)$$

where K_{CM} is the Clausius-Mossotti factor, defined as

$$K_{CM} = \frac{\varepsilon_p^* - \varepsilon_m^*}{\varepsilon_p^* + 2\varepsilon_m^*}, \qquad (1\text{-}7)$$

where ε_p^* and ε_m^* are the complex permittivity of the particle and the solution, respectively.

$$\varepsilon_p^* = \varepsilon_p - j\frac{\sigma_p}{\omega}, \qquad (1\text{-}8)$$

$$\varepsilon_m^* = \varepsilon_m - j\frac{\sigma_m}{\omega}, \qquad (1\text{-}9)$$

where ε_p and σ_p are the permittivity and conductivity of particle, respectively; ε_m and σ_m are the permittivity and conductivity of solution, respectively; and ω is the angular frequency of the electric field.

Therefore, the DEP force can be expressed as:

$$\overrightarrow{F_{DEP}} = 4\pi r_p^3 \, \varepsilon_m \cdot K_{CM} \cdot (E \cdot \nabla) \, \vec{E} = 2\pi r_p^3 \, \varepsilon_m \, \mathrm{Re}[K_{CM}] \nabla \vec{E}^2. \tag{1-10}$$

It can be seen from Equation (1-7) that the K_{CM} coefficient are related to the complex permittivity of solution and particle and the frequency of electrical signal, as shown in Figure 1.10.

When $\mathrm{Re}[K_{CM}] > 0$, the DEP force is positive DEP (pDEP), and under the pDEP force, the particle will be attracted to the region with larger gradient of electric field squared; when $\mathrm{Re}[K_{CM}] < 0$, the DEP force is negative DEP (nDEP), and under the nDEP force, the particle will be pushed to the region with smaller gradient of electric field squared; when $\mathrm{Re}[K_{CM}] = 0$, the DEP force is 0, and the corresponding frequency of electrical signal is the crossover frequency.

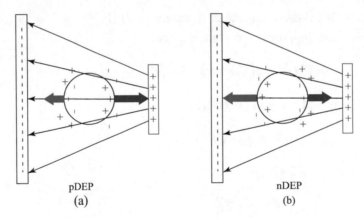

pDEP   nDEP
(a)   (b)

Figure 1.10: Schematic diagram of positive and negative DEP: (a) positive DEP; and (b) negative DEP.

When the particle is in a rotational electric field, since the dipole generation takes certain time, there is a phase shift θ between the direction of the dipole moment p and the direction of the electric field E, and the dipole moment is rotated parallel to the direction of the electric field, so the dipole moment p will continuously rotate with the electric field.

The average torque in a rotational electric field is the DEP torque [103]:

$$\Gamma_{ROT} = -\tfrac{1}{2} \, \mathrm{Re}[p \times \vec{E}]. \tag{1-11}$$

Substituting Equation (1-6) into (1-11) yields

$$\Gamma_{ROT} = -4\pi \varepsilon_m \, r_p^3 \, \mathrm{Im}[K_{CM}] \, E^2. \tag{1-12}$$

It can be known from Equation (1-12) that the rotational torque is related to the imaginary part of the K_{CM} coefficient, to the composite permittivity of the solution and particle and the electrical signal parameter. The positive and negative parts of Im[K_{CM}] determine the direction of the DEP torque. When the sign changes, the direction of the torque acting on the particles also changes.

1.3.2   DEP PARAMETER ANALYSIS

From Equations (1-10) and (1-12), it is known that DEP force and torque are related to size, electrical parameters of particle, electrical parameters of solution, and electric field.

    a. DEP force is related to the non-uniformity of the electric field, and the particles are subjected to DEP force only in a non-uniform electric field.

    b. DEP force is related to the particle size. The larger the particle, the greater the DEP force.

    c. The direction of DEP force is related to the real part of K_{CM} factor. When Re[K_{CM}]> 0, the DEP force is positive, that is, pDEP force, and the particle will move to the region with a large electric field gradient, when $Re[K_{CM}] < 0$. The DEP force is negative, that is, the nDEP force, and the particle will move to a region with a small electric field gradient.

    d. The DEP torque is related to the non-uniformity of the applied electric field; the particle is subjected to DEP torque only in a rotational electric field.

    e. The strength of DEP torque is related to the particle size. The larger the particle, the larger DEP torque.

    f. The direction of DEP torque is determined by the imaginary part of the K_{CM} factor. When the value of Im[K_{CM}] changes positively and negatively, the direction of rotation of the particle changes.

1.3.3   ADVANCES IN DEP-BASED SINGLE-CELL MANIPULATION

With the development of MEMS technology, the application range of DEP technology has been extensively expanded. Researchers have proposed various types of DEP microfluidic chips that can be applied to biological objects of different scales, ranging from manipulation of DNA [104], viruses [105], bacteria [106, 107], cells [108, 109], particles [110], and even model organisms.

Single-cell analysis is important for life sciences, clinical diagnosis, and drug evaluation [125, 126]. Traditional cell analysis methods collect the response of cell populations and can't accurately detect the information of individual cells. Biologists have confirmed that even mutated genes carried by CTCs derived from the same tumor are different [127]. In addition, a small number of cells may have profound effects on cancer origin, progression, and therapeutic research [128]. Therefore, single-cell analysis techniques are essential, so the corresponding single-cell sample manipulation is also critical to the analysis.

Among many techniques, DEP has been widely used in single-cell manipulation because of its label-free, non-invasive and highly selective properties [129]. DEP is less harmful to cells, and the viability of biological cells after DEP can be as high as 91.9% [130]. At present, DEP manipulations on single cells can be mainly divided into single-cell capture [131], single-cell translation [132], cell fusion [133], cell rotation [134], etc. For example, Wu et al. proposed a single-cell capture chip with microelectrode arrays, in which two electrodes inside the well can generate pDEP force to attract single cells into the wells (Figure 1.11(a)) [135]. Şen et al. proposed a single-cell capture chip with 900 gourd-shaped microwell arrays (Figure 1.11(b)) [136]. The electrodes in the arrays interact with the top ITO electrode to capture different types of single cells. Then, the captured cells are paired by nDEP force.

Gel et al. designed a microfluidic chip for cell fusion (Figure 1.11(c)) [137]. The microstructure in the microchannel is used to change the electric field distribution, and the cells are captured on both sides of the microstructure, and then the pulse is used to realize the electric fusion of the cells. In addition to the use of DEP force to capture single cells, single-cell rotation can be achieved by using DEP torque. For example, Bahrieh et al. used four planar electrodes to achieve single-cell rotation (Figure 1.11(d)) [138], by measuring cell rotation spectra after treatment with different drugs, electrical parameters of single cells can be measured for the study of the effects of drug on cells.

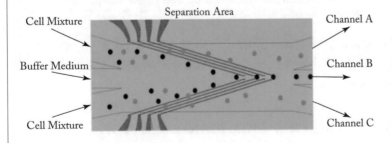

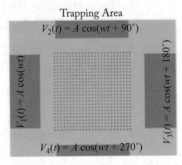

(a)

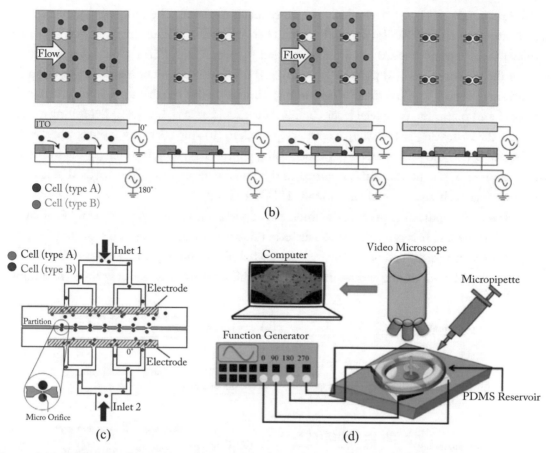

Figure 1.11: DEP microfluidic chip: (a) single-cell DEP capture chip [135], based on and used with permission from the American Chemical Society; (b) single-cell DEP capture pairing chip [136], based on and used with permission from the Royal Society of Chemistry; (c) cell electro-fusion chip [137], based on and used with permission from AIP Publishing; and (d) single-cell electro-rotation chip [138], based on and used with permission from the Royal Society of Chemistry.

1.3.4   ELECTRODE FABRICATION OF DEP CHIPS

At present, 2D planar electrodes have been widely used in microfluidic chips, but 2D planar electrode structure has limitations: the electric field generated by the 2D planar electrode decays quickly; the effective working regions are small; and the electric field is non-uniform distributed in the vertical direction. However, the thick electrodes can break through the limitations of the 2D planar electrodes, and the electric field generated by the thick electrodes can maintain uniformity in the vertical direction, increase the spatial distribution range of the non-uniform electric field, and enlarge the working regions of DEP force.

The materials of planar electrodes are mainly metal materials. The processing generally includes photolithography, thin film deposition, and chemical etching. There are two types of planar electrodes processing methods. (a) Sputter deposition (Figure 1.12(a)): (i) coating a layer of photoresist on the substrate, and (ii) photolithography, (iii) depositing a layer of metal on the substrate by a sputtering process, and (iv) finally degumming. This method is suitable for electrode materials such as gold and platinum. (b) Metal layer chemical etching (Figure 1.12(b)): (i) depositing a layer of metal on the insulating substrate by a sputtering process, (ii) spin coating a layer of photoresist on the substrate, (iii) the photoresist is patterned using photolithography, (iv) etching the substrate, the metal covered by the photoresist is retained, and (v) finally degumming. This method is suitable for electrode materials such as indium tin oxide (ITO) and copper.

In view of the sputtering process, the thickness of planar electrode is from several nanometers to several micrometers. It is not suitable to fabricate thick electrodes of several ten micrometers or even several hundred micrometers. The main reason is that the sputtering speed is generally 100 Å/min, and the use of a sputtering process for forming a thick electrode is not only time consuming but also high cost.

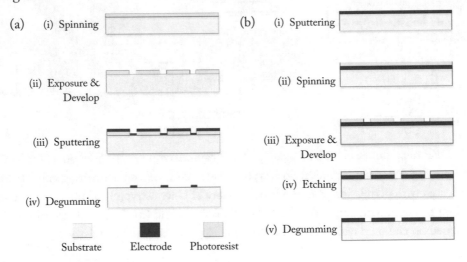

Figure 1.12: 2D planar electrodes processing methods: (a) sputter deposition; and (b) metal layer chemical etching.

The fabrication methods for thick electrodes currently used in microfluidics can be divided into the following.

1. Machining method, cutting a metal foil to form a thick electrode by machining, and embedding the electrode into PDMS microchannel. Zeinali et al. used mechanical processing to make thick electrodes and embedded them in the microchannel to achieve cell sorting [139]. However, machining can only be done in a few hundred

micrometers. It is difficult to achieve a finer electrode structure, and the surface of the electrode is rough.

2. Electroplating method, by depositing a metal on a 3D structure or depositing a metal on a substrate having a 3D structure to form a thick electrode. Voldman et al. used a plating method to fabricate a cylindrical 3D electrode [140], as shown in Figure 1.13 (a). The capture and release of single cells is achieved by the DEP force generated by the 3D microelectrodes. However, the fabrication process is complex and high cost.

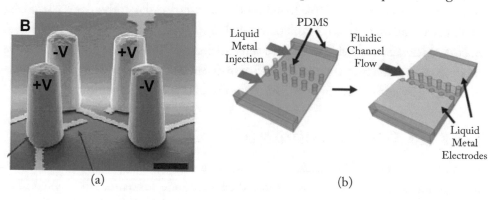

Figure 1.13: Microfluidic on-chip thick-electrode fabrication method: (a) electroplating [140], used with permission from the American Chemical Society; and(b) injection method [141], based on and used with permission from the Royal Society of Chemistry.

3. Injection method, by solidifying a liquid or semi-solid conductive material into a microchannel to form thick electrodes. There are two types of structures according to whether the electrodes are in contact with solution. One type is that the electrodes contact with the solution. So et al. injected liquid metal through a syringe to make thick electrodes as shown in Figure 1.13(b) [141]. Although this method can well embed thick electrodes in the micro-channel, it needs to ensure good injection precision such that the conductive silver glue cures quickly. The other type is that the electrodes don't contact with the solution [142, 143]. And the form of DEP is called cDEP (contactless-DEP). Shafiee et al. designed a cDEP microchannel structure for cell enrichment [144]. Electrolyte is used as electrodes in the electrode microchannel, and electrical signal is applied to the electrolyte to generate a DEP force for cell enrichment. This method is easier to prepare electrodes, but the biggest problem is that the electrodes don't contact with the solution, and a high voltage is required to overcome the influence of the insulating layer between the electrodes and the cell

solution. For example, an electric field of 25 V/μm is required to overcome the influence of PDMS insulation layer.

4. Photolithography mold method, a nano-conductive material (nano-carbon powder or nano-silver powder) is mixed with PDMS, and the mixture is plastered to a photolithography mold, and then use blade to scrape and form a pattern, which is embedded into the PDMS microchannel [145]. Lewpiriyawong et al. prepared a thick-electrode chip using a mixture of nano-silver powder and PDMS (AgPDMS) to achieve separation of particles and cells [146]. Based on the same processing technology, Marchalot et al. used a mixture of nano-conductive carbon powder and PDMS (C-PDMS) to make a thick-electrode chip to achieve cell enrichment [147]. This PDMS conductive mixture retains the properties of PDMS, which can be easily bonded to the substrate. The thick electrodes have high conductivity, ductility and repeatability.

1.4   RESEARCH PURPOSES AND SIGNIFICANCES

Biomanipulation based on DEP technology is currently a research hotspot in academia. In order to overcome the limitations of 2D electrodes, the thick-electrode structures are adopted to design single-cell operation chip, especially to solve the problem of single-cell 3D rotation. First, by constructing a multi-electrode structure with thick electrodes to realize single-cell 3D electro-rotation, electrical parameter measurement and morphology imaging. This work can demonstrate the value of thick-electrode DEP multi-electrode structure in the field of single-cell manipulation. On this basis, the thick-electrode DEP multi-electrode structure is expanded by opto-electronic integration, and dual optical fibers are embedded in the thick electrodes, which enable the chip to rotate and stretch single cells, achieving multi-parameter measurement of mechanical and electrical properties of single cells, extends the application of thick-electrode DEP in biological manipulation and analysis.

1.5   MAIN CONTENT OF THE BOOK

Chapter 1: Introduction. The background and significance of the topic. First, the background and current situation of microfluidic development are briefly summarized. The materials and processing methods of the microfluidic chip were analyzed, and the typical biological manipulation mechanisms are compared according to the technical means.

The research progress of single-cell manipulation by DEP is introduced. Then the research status of electrode processing in DEP technology is analyzed, and the method basis is provided for the design of chip structure. Finally, the purpose and significance of this research are proposed according to the requirements of biological applications.

Chapter 2: Thick-electrode DEP for single-cell 3D rotation. A thick-electrode DEP multi-electrode chip was proposed for single-cell 3D electro-rotation. First, the thick-electrode multi-electrode chip structure is presented. The working principle and design idea of the chip are introduced, and the feasibility of 3D rotation structure based on multiple thick electrodes is verified by simulation analysis, then set up the experimental platform. Based on the rotation platform, 3D electrorotation of single cells is demonstrated, and single-cell electrical parameters and physical topography parameters are measured. The work revolves around the application of multi-electrode structure of thick-electrode DEP in single-cell 3D rotation, which extends the application of thick-electrode DEP.

Chapter 3: Opto-electronic integration of thick-electrode DEP microfluidic chip. By extending the function of the multi-electrode chip with thick-electrode DEP, the fiber optical stretcher is embedded into the thick electrodes to stretch and rotate the single cell. First, the application requirement of single-cell multi-parameter measurement for thick-electrode DEP optoelectronic integrated chip is introduced. Then, the principle of single-cell optical stretcher and the analysis method of mechanical properties are analyzed. Subsequently, the thick-electrode DEP opto-electronic integrated chip is presented, which achieves mechanical and electrical property measurements of single cells. The work of this chapter focuses on the application of thick-electrode DEP opto-electronic integrated structure in single-cell multi-parameter measurement, and further broadens the application of thick-electrode DEP.

Chapter 4: Summary.

CHAPTER 2

Thick-Electrode DEP for Single-Cell 3D Rotation

2.1 INTRODUCTION

At present, most of the DEP chips are based on 2D planar electrodes, which have problems such as fast electric field attenuation and small working distance. To solve the limitations, thick electrodes have been widely used in microfluidics in recent years because of its large working areas and uniformity of electric field in the vertical direction [148, 149].

With the development of MEMS technology, researchers have proposed a variety of thick-electrode microfluidic chips with different functions. For example, in cell-sorting applications, Wang et al. [150] designed a thick-electrode microfluidic chip to achieve cell sorting (Figure 2.1(a)). The thick electrodes are deposited on the SU-8 microstructure by electrodeposition, but the fabrication is complicated, time-consuming, and costly. Li et al. used a method of heating low-melting metal microspheres in a microchannel to make thick electrodes and used them for cell sorting [151]. However, this method has insufficient processing precision, and the shape of the metal after melting is less controllable, and it is easy to overflow the flow channel. Thick-electrode DEP has also been used for particle separation. For example, Kang et al. designed a 3D embedded electrodes microfluidic chips for particle separation (Figure 2.1(b)) [152].

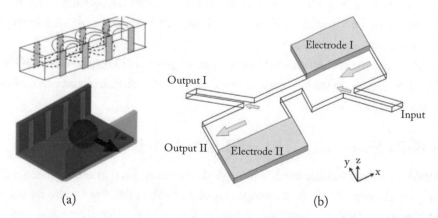

(a)   (b)

Figure 2.1: Microfluidic chips based on thick electrodes: (a) a thick-electrode microfluidic chip for cell sorting [150]; and (b) A thick-electrode microfluidic chip for particle separation [152], based on and used with permission from Elsevier.

The application of thick electrodes in microfluidic chips has demonstrated the potential of thick-electrode DEP in microfluidics. However, the fabrication process of existing thick-electrode DEP chips is rather complicated, and the electrode structure is relatively simple, which limits the application of the thick electrodes.

The basic thick-electrode DEP composed of two thick electrodes can generate DEP force which can translate single cells. However, it is necessary to extend the basic two-electrode structure to multiple electrodes to achieve more complex manipulations, such as single-cell rotation. In this work, a thick-electrode DEP chip for single-cell 3D rotation was proposed by using carbon black-PDMS (C-PDMS, a mixture of nano-conductive carbon powder and PDMS) as electrode material, which expands the application prospect of thick-electrode DEP in microfluidics.

Single-cell 3D rotation means that cells suspended in solution can rotate about X/Y/Z-axis and play an irreplaceable role in single-cell analysis. For example, when analyzing the biophysical properties of cells, it is necessary to perform 3D surface imaging of cells [153], and even internal structure scanning [154]. In order to obtain accurate imaging results, it is essential to rotate the cells about more than one axis to obtain multi-dimensional image sequences, then reconstruct 3D model. But 3D rotation is not as easy to implement; most mammalian cells are 10–100 μm in diameter, and easily sink in solution. Although the current cell rotation can be achieved by various methods, such as mechanical, optical, magnetic, acoustic, or electrical means, most of the methods can only achieve in-plane rotation. Even for several existing 3D rotation methods, there is a problem that the rotation control is unstable.

Cells are polarized in a rotational electric field and rotated by the torque generated by electric field. The electrical parameters of the cells such as cell membrane capacitance and cytoplasmic conductivity of the cells can be measured by analyzing the rotation spectrum. However, at present most electro-rotation methods using planar electrodes cannot achieve 3D rotation. And the rotation speeds are different when cells are in different positions, which makes it impossible to accurately measure electrical parameters of the cells.

This chapter presents an "Armillary Sphere" type single-cell 3D rotation chip. 3D rotation is realized by the thick-electrode multi-electrode structure, and the electrical and physical properties are measured based on 3D rotation.

2.2   PROGRESS IN CELL ROTATION MANIPULATION

Single-cell rotation has been widely used in biological operations such as cell injection, cell nuclear extraction, and cell cloning. With the development of MEMS technology, there have been many reports on the methods of cell rotation such as mechanical, acoustic, electric, optical, and magnetic.

1. **Mechanical methods.** Generally, a suction tube and an injection needle are used for operation. First, the cells are adsorbed by the suction tube, and then the cells are ro-

tated by the needle. Chen et al. used this method to achieve rotation and enucleation of bovine oocytes, as shown in Figure 2.2(a) [155]. However, the rotation efficiency of the mechanical methods is low, and a precise positioning manipulator is required to accurately control the displacement of the glass needle. Furthermore, mechanical methods are contact operation, which are easy to cause damage to the cells.

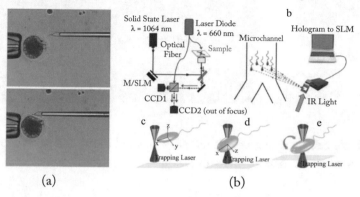

Figure 2.2: (a) Mechanical method for single-cell rotation [155]; and (b) sperm cell rotation based on the optical method [157], based on and used with permission from the Royal Society of Chemistry.

2. **Optical methods.** The methods use a laser beam to generate axial and gradient forces for cell capture and precise control of cell movement [156]. However, it is difficult to achieve cell rotation in one laser beam. Merola et al. proposed a single-cell rotation microdevice based on an unstable Gaussian beam to achieve sperm cell rotation [157], and 3D imaging of sperm by digital holography, as shown in Figure 2.2(b). But the rotation method requires expensive optical systems with complex and precise controls.

3. **Magnetic methods.** The use of magnetic field to control cells requires modification of the cell surface with micro-magnetic beads and the external magnetic field to achieve cell movement [158, 159]. Elbez et al. used an external 3D magnetic coil to generate a 3D magnetic field to achieve 3D rotation of the cells [158], as shown in Figure 2.3(a). However, the external magnetic coil has a large volume, and the fabrication are complicated.

4. **Acoustic methods.** The transducer vibrates the liquid in the microchannel and creates vortexes that allow single cells or even organisms (such as *Caenorhabditis elegans*) to rotate [160, 161]. Tony Huang group developed a surface acoustic wave rotation structure. As shown in Figure 2.3(b), the shapes of the vortexes were changed by adjusting electric parameters of piezoelectric transducer. However, acoustic methods

don't achieve precise and stable control of single cell rotation, and the states of rotation depend on the shape of the bubbles and structures.

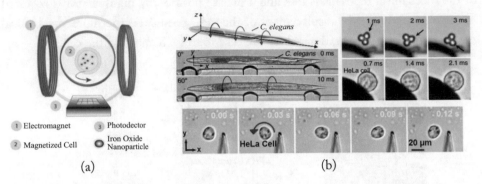

(a)   (b)

Figure 2.3: (a) The single-cell rotation based on magnetic method [158]; and (b) the rotation of C. elegans and single cells based on acoustic methods [160].

5. **Fluidic methods.** Setting structures with specific shape in the microchannel produce vortexes to make cells rotate [162]. The shape of structure can be either a microcolumn or a groove. Patrick Shelby et al. designed a horizontal groove structure that produces horizontally rotational vortexes to achieve in-plane rotation of single cells [163], as shown in Figure 2.4. Shetty et al. designed a vertical groove structure that produces a vertically rotational vortex to achieve out-of-plane rotation of single cells and 3D imaging of cells [164]. The fluidic methods have low cost and simple operation, and don't require any pre-treatment on the cells, but the single-cell loading efficiency is low, and only 1D rotation can be achieved.

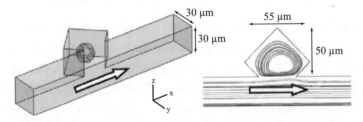

Figure 2.4: Single-cell rotation based on fluid method; horizontal single-cell rotation using horizontal concave microchannels [163], based on and used with permission from the Royal Society of Chemistry.

6. **DEP methods.** If cell is polarized under a rotational electric field, it will be subjected to DEP torque. A common device utilizes four planar electrodes to apply ac signal with different phase shift on the four electrodes to form a rotational electric field.

However, the spatial position of the cell is unstable, and the electric field decays rapidly in the vertical direction.

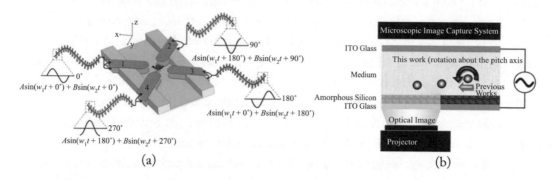

Figure 2.5: Single-cell rotation based on DEP: (a) single-cell rotation using octagonal planar electrodes [165], based on and used with permission from the Royal Society of Chemistry; and (b) single-cell rolling operation using light-induced DEP [166], based on and used with permission from AIP Publishing.

In order to overcome the decay problem of electric field, Han et al. designed two sets of upper and lower planar electrodes. The octagonal electrodes ensure the stable positioning of the cells and achieve stable rotation of the cells [165], as shown in Figure 2.5(a). Liang et al. used an optoelectronic material to prepare virtual electrodes, and the distribution of electric field was changed by adjusting the position and shape of light [166], as shown in Figure 2.5(b). Cells at the edge of the virtual electrodes are vertically rotated under the DEP torque, but the area of the vertical rotation is limited to the boundary of the virtual electrodes.

In summary, a lot of methods base on different technologies can achieve cell rotation, and have their own advantages and applications. However, most of the methods can only achieve 2D/1D cell rotation. Cell rotation based on DEP technology enables precise rotation control and can be used for measurement of cellular electrical parameters. The work of this chapter focuses on how to construct a thick-electrode DEP multi-electrode chip to realize the operation and analysis of single-cell 3D rotation.

2.3   THICK-ELECTRODE MULTI-ELECTRODE CHIP DESIGN

2.3.1   PRINCIPLE AND DESIGN OF THICK-ELECTRODE MULTI-ELECTRODE CONSTRUCTION

Based on the theory of DEP, how to design the electrode structure to generate a 3D rotational electric field is the key to realizing 3D rotation. In our previous work, a microchip with four thick electrodes and two transparent planar electrodes can successfully obtain single-cell 3D electro-rotation [167]. However, the device is an open chip and prepared by a machining process. It has a problem in loading single cells, and is suitable only for cells with large size such as bovine oocytes. Moreover, the single-cell rotation is not stable, and the sinking problem easily occurs.

The use of thick electrodes in rotation applications has many advantages: (a) thick electrodes can overcome the limitations of 2D planar electrodes that can't achieve 3D rotation; (b) thick electrodes can solve the shortcoming of the electric field attenuation of the planar electrodes in the vertical direction, greatly increasing the rotatable spatial region; and (c) the electric field distribution of thick electrodes in the vertical direction is more uniform and the cell rotation is more stable, which ensures the accuracy of the results for electrical measurements.

Figure 2.6 shows a schematic diagram of a thick-electrode multi-electrode DEP rotation structure. The multi-electrode structure consists four thick electrodes and a bottom electrode which form an open electrode chamber.

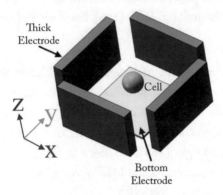

Figure 2.6: 3D rotation electrode chamber structure.

Rotational electric fields in horizontal and vertical direction are generated by applying different electrical signal configuration on different electrodes. Applying signals with the same amplitudes, the same frequencies but different phase shifts on four vertical electrodes can form a horizontally rotational electric field in the chamber. As shown in Figure 2.7(a), the cells are subject to torque in a horizontally rotational electric field. Applying signals with the same amplitudes, same frequencies, but different phase shifts on the opposite two vertical thick and bottom electrodes, as shown in Figure 2.7(b), can generate a vertical rotational electric field in the chamber.

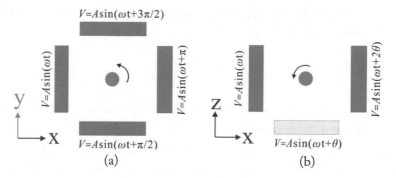

(a)   (b)

Figure 2.7: Signal configuration for 3D rotation: (a) in-plane rotation and (b) out-of-plane rotation.

Through simulation, the electric field in the chamber can rotate horizontally and vertically under different electrical signal configurations. Figure 2.8(a) is a horizontal rotation simulation diagram. Applying signals with phase shift of 90° (Vp-p = 10 V, f = 1 MHz) on four thick electrodes will form a horizontally rotational electric field in the chamber. A cell can do in-plane rotation by a clockwise DEP torque under the horizontally rotational electric field. Figure 2.8(b) is a vertical rotation simulation diagram. Applying signals with phase shift of 120° (Vp-p = 10 V, f = 1 MHz) on two opposite thick electrodes and bottom electrode. A vertical rotational electric field is formed in the vertical direction; the cell is subjected to torque for out-of-plane rotation. Since the electrodes are not completely symmetrically distributed in the vertical direction, the strengths of the vertical rotational electric field are not the same at different times in one cycle.

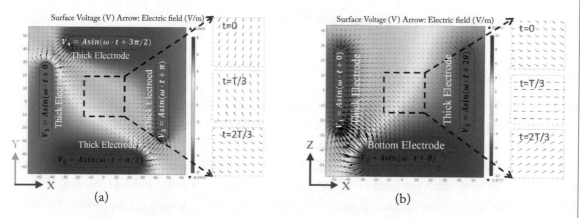

(a)   (b)

Figure 2.8: Simulation of a 3D rotational electric field: (a) top view of the electrode chamber (in-plane rotation); and (b) side view of the electrode chamber (out-of-plane rotation).

2.3.2   DESIGN AND SIMULATION OF 3D ROTATIONAL STRUCTURE OF "ARMILLARY SPHERE"

Chip structure

Although the above-mentioned open structure can realize 3D rotation of single cells, single-cell loading is nontrivial. In order to improve the single-cell loading efficiency, a microchip combining the thick electrodes and microchannel is proposed, as shown in Figure 2.9. The materials of four vertical thick electrodes are C-PDMS, and the bottom electrode is transparent electrode ITO. For single-cell loading, a V-shape single-cell capture structure is designed in the microchannel.

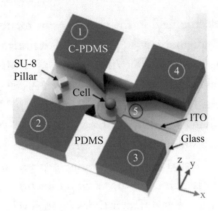

Figure 2.9: Schematic diagram of the single-cell 3D rotational chip structure.

Working procedures

Figure 2.10 shows the working procedures: (a) inject the cell solution and capture one single cell using the V-shape pillars as the trap site; (b) using a back flow to release the captured cell from the trap site, and when the cell flow to the electrode chamber, an electrical signal can be applied to the two thick electrodes. When the resulting nDEP force and Stokes force are balanced, the cell is kept in the chamber; (c-e) after the position of the cell in the chamber is stabilized, 3D rotation of the cell is achieved by applying different signal configurations on the electrodes; (f) recovering the cell by flushing the flow and stopping the electrical signals.

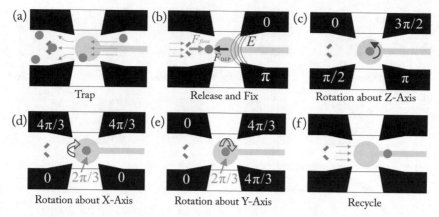

Figure 2.10: Chip working procedure diagram.

Electrode shape design

The DEP force and torque are proportional to ∇E^2 and E^2, respectively, and are related to the shape of the electrode. Figure 2.11(a–c) shows three typical electrode geometries (circular, square, and sharped) with the same microchannel width (200 μm). $V_1 = 10\sin(\omega t)$, $V_2 = 10\sin(\omega t+\pi/2)$, $V_3 = 10\sin(\omega t+\pi)$, and $V_4 = 10\sin(\omega t+3\pi/2)$ are applied to the four thick electrodes, and the bottom electrode is floating. Assuming $\varepsilon_m = 100\varepsilon_0$, $R_{cell} = 6$ μm, $\mathrm{Re}[K_{CM}] = 0.5$, Figure 2.11(d) is the average strength distribution of the electric field over a period along the line A-A corresponding to the three structures. The electric field generated by the sharped electrode has the highest average strength, which means the generated torque is the largest.

The DEP force generated by the two electrodes is analyzed. $V_3 = 10\sin(\omega t)$, $V_4 = 10\sin(\omega t+\pi)$ are applied to the electrodes 3, 4, and the remaining two electrodes and the bottom electrode are floating. Figure 2.11(e) shows the distribution of DEP forces along the cut line A-A. It can be found that the sharp electrode produces the greatest DEP force, approximately twice that of the circular and square electrodes. Thus, the sharp electrode is selected as the electrode geometry.

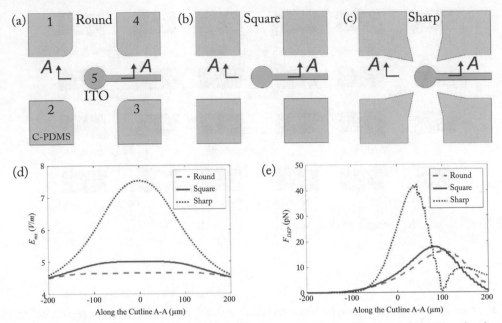

Figure 2.11: Comparison of electric field and DEP force of different shapes of electrodes: (a–c) top view of electrode with three typical shapes; (d) electric field strength along the line *A-A* (Vp-p = 10 V); and (e) along the line *A-A* The distribution of the DEP force on (Vp-p = 10 V).

Single-cell V-shape capture structure design

The V-shape trap structure placed in the microchannel is used to capture a single cell. Under the laminar flow, a single cell will be trapped at the V-shape trap structure. Once a single cell is captured, the flow resistance at the V-shaped trap structure will increase and the remaining cells will flow away from both sides. For the captured cell, back flow is used to push it away from the trap site and transfer it to the electrode chamber.

The microchannel size is 200 μm×160 μm wide, the V-shape structure is 15 μm × 15 μm × 25 μm, and the gap of V-shape is 10 μm. Because the cells are soft, the cell at the trap site would be deformed easily by the flow force and squeezed out of the gap. On the other hand, it is necessary to avoid too much shear force generated by the fluid to affect the cell. The shear force of the flow acting on the cell is estimated to be 0.076 dyne/cm² at a flow rate of 20 μm/s. According to the literature, the effect of this shear force on the cell can be negligible.

Single-cell capture positioning simulation

After single-cell capture, the cell is transferred to the electrode chamber using back flow, and the location of the cell in the chamber can be determined in two ways. The first method uses visual feedback to observe the position of the cell in real time, and stops the micropump immediately when the cell moves into the chamber. However, this method has delay time and the cell easily flows out of the electrode chamber. The second method uses the DEP force to balance the stokes force. When back flow pushes the trap cell into the electrode chamber, an electrical signal is applied to the two thick electrodes on the right side, which creates a DEP force between the two electrodes. Figure 2.12(a) is a simulation of the electric field strength (Vp-p = 10 V, f = 100 kHz). Figure 2.12(b) shows the distribution of DEP forces along the A-A cutline at different voltage amplitudes. Assuming cell flow rate 40 μm/s, the stokes force is about 24 pN. It is possible to estimate the position at which the DEP force equal to stokes force. The larger the voltage amplitude, the more the balance position is to the left. For example, when the voltage amplitude is 10 V, the balance position of the cell is approximately at the center of the chamber.

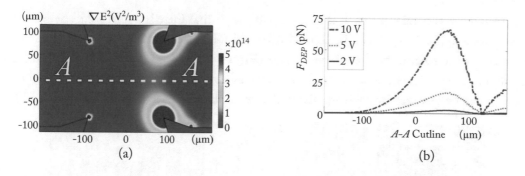

(a)   (b)

Figure 2.12: Simulation results of DEP force: (a) simulation of electric field strength; and (b) the distribution of DEP force on A-A cutline.

3D rotational electric field simulation

Horizontal rotation

Applying the same amplitude and frequency electrical signals (Vp-p = 10 V, f = 1 MHz) with a phase shift of 90° on the four C-PDMS thick electrodes can generate a horizontally rotational electric field in the chamber, causing the cell to do in-plane rotation. Figure 2.13 shows the electric field distribution simulation of the electrode chamber. Figure 2.13(a) shows that the electric field in the chamber rotates clockwise in one signal period. Figure 2.13(b) shows the electric field in the chamber rotates counterclockwise in one signal period once reversing the signals to the four electrodes.

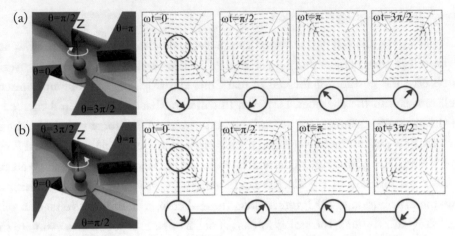

Figure 2.13: Simulation of horizontal rotational electric field: (a) clockwise rotation about the Z-axis; and (b) counterclockwise rotation about the Z-axis.

Vertical rotation

In order to achieve vertical rotation, it is necessary to ensure that there is a phase shift between the electrical signals applied to the two thick electrodes and the bottom electrode. Figure 2.14 shows the variation of the electric field at different time (Vp-p = 10 V, f = 1 MHz). In one period, the electric field in the vertical direction rotates clockwise, as shown in Figure 2.14(a). The direction of cell rotation can be changed by adjusting the sequence of phase shift, as shown in Figure 2.14(b).

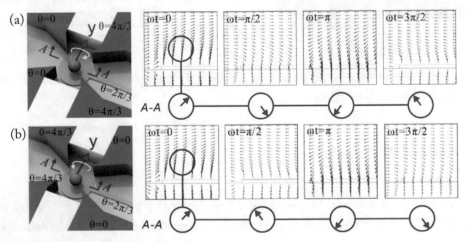

Figure 2.14: Simulation of vertical rotational electric field (about Y-axis).

Changing the electrical signal configuration on the electrodes enables the electric field in the chamber to rotate about the X-axis. Figure 2.15 shows the change in electric field at different time (Vp-p = 10 V, f = 1 MHz). In one period, the electric field in the vertical direction rotates counterclockwise, as shown in Figure 2.15(a). When the phase shift is switched to 240°-120°-0°, the electric field rotation can be changed to clockwise, as shown in Figure 2.15(b).

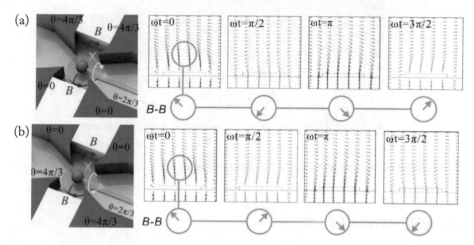

Figure 2.15: Simulation of vertical rotational electric field (about X-axis).

Overcome the cell sinking problem

Since the cell density is generally slightly greater than that of the solution, cells will sink to the bottom. Once cells sink down to the bottom, they will adhere to the electrode surface, causing a friction between cells and the bottom, affecting the rotation state. When the DEP torque acting on the cells does not overcome the friction between the cells and the bottom, cell rotation can't be achieved. Therefore, solving the problem of cell sinking is a prerequisite for ensuring a stable rotation of cells in the vertical and horizontal directions. The electrodes could produce a vertically upward DEP force by applying a signal with a phase shift of 180° (Vp-p = 10 V, f = 100 kHz) to two thick electrodes and bottom electrode. Figure 2.16 shows the simulation result of the electric field distribution. There is an electric field gradient in the vertical direction, and the direction of generated nDEP force is upward. When the sum of the DEP force and the buoyancy of the cell is equal to gravity of the cell, it can prevent the cell from sinking:

$$F_{DEP} + F_{bouyancy} = G. \tag{2-1}$$

As the DEP force decreases with increasing distance in the vertical direction, there would exist an equilibrium position for the cell in the vertical direction. The strength of DEP force can be adjusted by changing the amplitude of the signal, which can regulate the equilibrium position of the cells in the vertical direction.

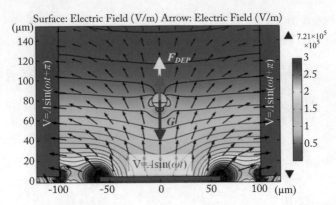

Figure 2.16: Simulation of cell suspension electric field strength.

2.4   CHIP FABRICATION

The chip is composed of upper and lower layers. The upper layer has a microchannel and four thick electrodes, and the lower layer has a V-shape trap structure and a bottom ITO electrode. Figure 2.17 is the mask of the upper layer (a) and the lower layer (b). In order to avoid electric short between electrodes, the thick electrodes can't make the microchannel closed at both ends, so four peripheral outlets need to be designed on the periphery of the microchannel, and the microchannel is sealed at both ends by injection glue or PDMS at the peripheral outlets.

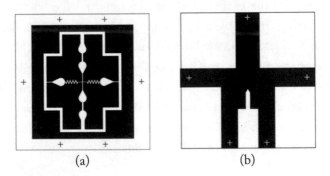

Figure 2.17: Mask design: (a) thick electrode design; and (b) bottom electrode design.

The fabrication procedure of the microdevice is shown in Figure 2.18.

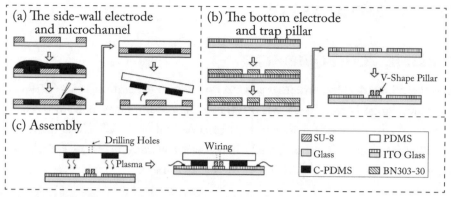

Figure 2.18: Chip fabrication: (a) thick electrode and microchannel fabrication: (i) A 160-μm thick SU-8 photoresist (SU-8 2075) is spin coated on a glass substrate. The photoresist is patterned using a photolithography process to form a SU-8 mold; (ii) press the C-PDMS onto the SU-8 mold and then slowly scrape off the excess C-PDMS with a blade. Then, it is baked at 80° C for 30 min for curing; (iii) pour the PDMS onto the cured C-PDMS mold and remove bubbles. Then, it is cured at 80° C for 30 min; (iv) the PDMS and C-PDMS are removed from the SU-8 mold; and (v) punching holes on PDMS/C-PDMS; (b) fabricate bottom electrode and V-shape structure: (i) spin coating and pattern 1–2-μm thick negative photoresist (BN303-30) on the ITO glass substrate; (ii) etching the ITO substrate to form bottom electrode; (iii) spin coating SU-8 photoresist of about 25-μm thickness on the ITO glass and fabricating a V-shape structure; and (c) chip assembly: (i) the upper layer and the bottom layer are treated by plasma-oxygen (power 30 W, time 60 s), and are aligned and bonded together; (ii) blocking of the microchannel. Since C-PDMS electrodes are independent of each other, the formed microchannel can't be completely closed, thus port sealing is required. To do so, a small amount of PDMS is injected into the four inlets, and baked at 80° C for 30 min for curing; and (iii) external wiring of the chip, the ITO electrode leads the C-PDMS electrode through the wire, and the inlet and outlet of the microchannel are led out through pipe.

Figure 2.19 shows a photo of the microchip.

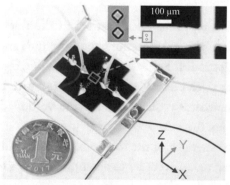

Figure 2.19: The photo of the 3D rotation microchip.

2.5   EXPERIMENTAL SETUP

2.5.1   EXPERIMENTAL EQUIPMENT

The main equipment includes: four-channel arbitrary waveform generator (TGA-12104, TTi) that the signal frequency range is from 0–25 MHz, and the amplitude range is from 0–20 Vp-p, micropump (Legato 200, KD Scientific), inverted microscope (Nikon Ti-U); CCD camera (acA640-120gm, Basler, 30–100 frames/s). A homemade Labview software controls the switching of signal configuration and real-time video acquisition of the CCD camera.

2.5.2   SIGNAL CONFIGURATION

Table 2.1 shows the signal configurations of each electrode in different working modes.

Table 2.1 Electrical signal configuration of cell loading and 3D rotation mode					
Electrodes	**1**	**2**	**3**	**4**	**Bottom**
Cell loading	Float	Float	0	π	Float
X-axis rotation (+)	0	$4\pi/3$	$4\pi/3$	0	$2\pi/3$
X-axis rotation (−)	$4\pi/3$	0	0	$4\pi/3$	$2\pi/3$
Y-axis rotation (+)	0	0	$4\pi/3$	$4\pi/3$	$2\pi/3$
Y-axis rotation (−)	$4\pi/3$	$4\pi/3$	0	0	$2\pi/3$
Z-axis rotation (+)	0	$\pi/2$	π	$3\pi/2$	Float
Z-axis rotation (−)	$3\pi/2$	π	$\pi/2$	0	Float

2.5.3   EXPERIMENTAL METHODS

Effect of electrical signal parameters

In order to generate a horizontal rotational electric field in the thick-electrode multi-electrode structure, it is only necessary to apply signal with the same amplitude and frequency, but with phase shift of 90° on the four thick electrodes. For out-of-plane rotation, the vertical rotational electric field needs to be formed by using thick electrodes in combination with the bottom electrode. Since the electrode structure is asymmetrically distributed in the vertical direction, the electric field strength in one signal period is not the same.

There is a difference in the strength of the rotational electric field generated by the different phase shifts in the out-of-rotation mode. In order to determine the optimal value of the phase shift, it can be considered from two aspects: (1) the average electric field strength in one signal period, and (2) the standard deviation of the electric field strength change in one signal period.

In order to quantify the above two factors, the center of the rotational chamber is selected as the calculation probing point. The phase shift of the signals on the electrodes is 0-θ-2θ (Vp-p = 10 V, f = 1 MHz), respectively. The simulation results show that when the phase shift is less than 30° or greater than 170°, a rotational electric field can't be generated. Figure 2.20(a) shows the corresponding curve of the rotational electric field strength when the phase shift changes from 30° to 170°. When the phase shift is 110°, the average electric field strength reaches the maximum. Figure 2.20(b) shows the standard deviation of the change in electric field strength in one period with different phase shifts. When the phase shift is 130°, the standard deviation of the change in electric field strength is minimized. Considering collectively, the value of phase shift is selected as 120°.

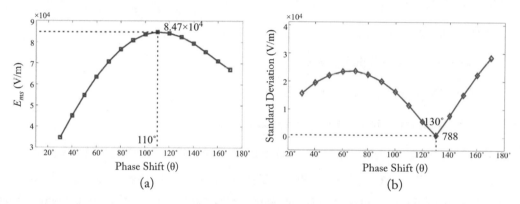

Figure 2.20: Out-of-plane rotation electrical signal configuration phase shift analysis: (a) phase shifts correspond to the electric field strength in one period; and (b) phase shifts correspond to the standard deviation of the electric field strength in one period.

DEP buffer solution

In order to minimize Joule heating, it is necessary to use a solution with low conductivity and to ensure that the solution is an isotonic solution. The DEP buffer formulation used is: 82% deionized water (w/v) +16% sucrose (w/v) + 2% PBS (w/v) with a conductivity of 36.5 mS/m.

2.5.4   3D ROTATION SPEED MEASUREMENT METHOD

Accurate measurement of cell rotation speed not only reflects the relationship between rotation speed and signal parameters, but also provides a basis for measurement of cellular electrical parameters. The speed measurement starts with using camera to collect the cell rotation motion, and then convert the rotation motion video files into frame sequences. One or more cell feature points are selected in the frame sequence, and the angle and rotation time corresponding to the rotation of the feature point for about one revolution are tracked. The angular velocity can be calculated from the angle/time.

2.6   SINGLE-CELL 3D ROTATION EXPERIMENT

2.6.1   CELL SAMPLE PREPARATION

Culture environment:

- cell culture incubator: Thermo LS-C0150; 37 ° C, 5% CO2 concentration.

- PBS (Corning, U.S.);

- trypsin (Corning, U.S.);

- fetal bovine serum (Sijiqing, China);

- DMEM (Thermo Fisher Scientific, U.S.);

- penicillin Streptomycin (Corning, U.S.); and

- cell culture medium: 90% DMEM medium, 9% fetal calf serum1%, double antibody.

Four types of cells—HeLa, B, HepaRG, C3H10—were used in this experiment. B cells are suspended cells, and the other three cells are adherent cells. For adherent cells, sample preparation steps are following. First, the medium, PBS and trypsin are heated to 37° C, then the medium is aspirated with a pipette. Subsequently, trypsin is added, after digesting for 10 minutes at 37°C, the cells change the adherent state to suspended. And then the same dose of medium is added to trypsin to stop the digestion process. Using pipette to transfer the suspended cells to a centrifuge tube, centrifuge at 1,000 r/min for 5 min. After centrifugation, aspirate the supernatant and add the medium to disperse the cells at the bottom of the centrifuge tube. DEP buffer is then added for use as an experimental sample.

Suspended cells are prepared by following the above steps for adherent cells, but after digestion.

2.6.2   CELL CAPTURE VALIDATION

The flow rate of cell solution (20 μm/s) was set in the experiment. At this flow rate, the cells are easily tracked under a bright field microscope. Under the action of laminar flow, the cells will move forward along the streamline. When the cells are in the center of the microchannel, they will stop at the V-shape structure. After one cell was captured, it was released from the trap site using a back flow. The flow rate of 40 μm/s was used to produce a Stokes force of ~24 pN. When the single cell was transported from the V-shape structure to the chamber, the nDEP force was generated by applying an electrical signal to the thick electrodes, and the cell stopped at an equilibrium position. Figure 2.21 shows a HeLa cell capture dynamic process.

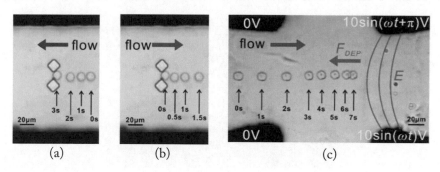

Figure 2.21: Single-cell capture, release, and trap experiments: (a) capture of one HeLa cell; (b) the captured cell release by back flow; and (c) cell was kept by nDEP force.

2.6.3   3D ROTATION EXPERIMENT

3D cell rotation about X/Y/Z-axis can be achieved by applying different electrical signal configurations.

(a) In-plane rotation mode

Figure 2.22 shows a HeLa cell rotating clockwise about the Z-axis at 270°/s (Vp-p = 6 V, f = 600 kHz). After changing the phase shift sequence of the four electrodes, the direction of rotation of the cell was reversed, and the cell rotated counterclockwise about Z-axis.

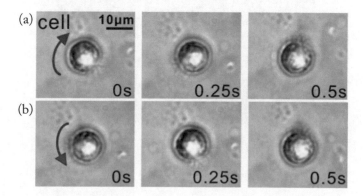

Figure 2.22: A HeLa cell rotates about the Z-axis (Vp-p = 6 V, f = 600 kHz).

(b) Out-of-plane rotation mode

Figure 2.23 shows a HeLa cell rotating about Y-axis (Vp-p = 6 V, f = 600 kHz). By changing the phase shift sequence of the electrical signals, the direction of rotation of the cells was changed.

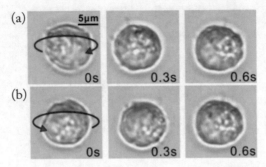

Figure 2.23: A HeLa cells rotated about the Y-axis (Vp-p = 6 V, f = 600 kHz).

(c) X/Y/Z axis rotation

Figure 2.24 shows a 3D rotation of a HeLa cell (Vp-p = 6 V, f = 600 kHz). First, it rotated about Z-axis, and then about X-axis, and finally about Y-axis.

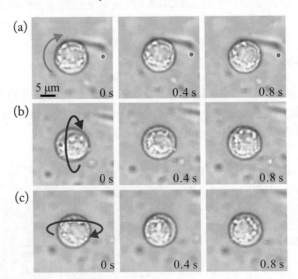

Figure 2.24: A HeLa cell rotated about X/Y/Z-axis.

(d) Different rotation modes about X-axis when cells are inside and outside the chamber

When electrodes are configured to rotate the cell about X-axis, the cell will have different rotation modes inside and outside the chamber (Figure 2.25).

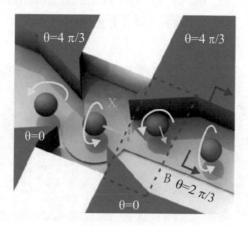

Figure 2.25: Cell rotation state in different regions when cell rotates about the X-axis

When the cell flows to the chamber and has not yet reached the chamber, at the boundary of bottom electrode, the cell first rotates about Z-axis. When the cell flows into the chamber, the rotation state of the cell is switched from in-plane rotation to out-of-plane rotation about X-axis. This result can be confirmed by simulation analysis. Under the configuration of the out-of-plane rotation about X-axis, outside the left boundary of the bottom electrode is in-plane rotational electric field (Figure 2.26(a)). Figure 2.26(b) shows that one B cell rotating about Z-axis and X-axis (Vp-p = 6 V, f = 600 kHz) across the boundary of bottom electrode.

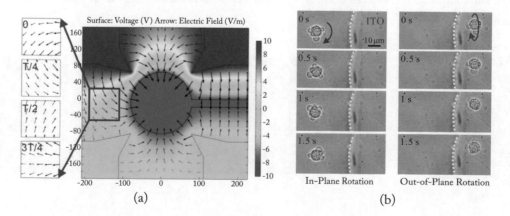

Figure 2.26: The rotation state of cells at the boundary of bottom electrode under the signal configuration for rotation about X-axis; (a) electric field simulation; and (b) B cell rotation experiment.

After exiting the electrode chamber, the cell can also rotate in the gap between the thick electrodes and the bottom electrode. If the cell is in a conical region (as shown by the dashed boxed area in Figure 2.25), it will make out-of-plane rotation about the axis perpendicular to the surface of the thick electrodes. Figure 2.27 shows that one HeLa cell adsorbed on the electrode surface under pDEP force (Vp-p = 6 V, f = 2 MHz) and doing out-of-plane rotation about the axis perpendicular to the electrode surface.

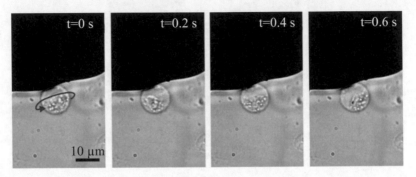

Figure 2.27: Bottom electrode edge cone area cell rotation state.

After escaping from the conical zone, between the bottom electrode and the thick electrode (as shown in Figure 2.25), the cells will resume rotation about the X-axis. There would be two rotation modes, depending on pDEP or nDEP signals applied. Under pDEP, the cell is adsorbed on the surface of the thick electrode and rotate about the X-axis. Figure 2.28(a) is the force model of single cell under pDEP force. Figure 2.28(b) shows that a HeLa cell adsorbed on the surface of the electrode (Vp-p = 6 V, f = 1 MHz) and rotating about the X-axis at speed of ~ 120°/s. Switching the phase shift sequence of the electrode signals, the cell rotation direction was switched to reverse rotation about the X-axis.

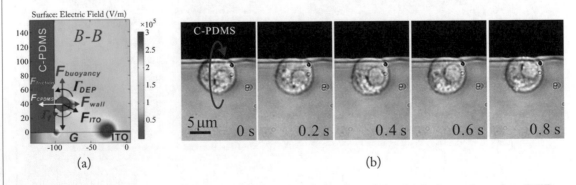

(a)   (b)

Figure 2.28: The rotation of cells between the bottom electrode and the thick electrode under pDEP force: (a) simulated force analysis; and (b) HeLa cells rotation around the X-axis under pDEP force.

Under nDEP, the cell between the two electrodes is repelled. Figure 2.29(a,b) is the force model and simulation.

The horizontal force analysis yields

$$F_{ITO} \cdot \sin \theta_1 = F_{C\text{-}PDMS} \cdot \sin \theta_2 . \qquad (2\text{-}2)$$

The DEP force generated by ITO and C-PDMS is calculated by Equation (1-10):

$$2\pi R_{cell}^3 \varepsilon_m \text{Re}[K_{CM}] \nabla E_{ITO}^2 \sin \theta_1 = 2\pi R_{cell}^3 \varepsilon_m \text{Re}[K_{CM}] \nabla E_{C\text{-}PDMS}^2 \sin \theta_2. \qquad (2\text{-}3)$$

$$\nabla E_{ITO}^2 \cdot \vec{x} - \nabla E_{C\text{-}PDMS}^2 \cdot \vec{x} = 0. \qquad (2\text{-}4)$$

The left side of the equation is actually the component ∇E^2 in the horizontal direction:

$$\nabla E^2 \cdot \vec{x} = 0. \qquad (2\text{-}5)$$

Also in the vertical direction, the cell is balanced:

$$F_{ITO} \cdot \cos \theta_1 + F_{C\text{-}PDMS} \cdot \cos \theta_2 + F_{buoyancy} = G, \qquad (2\text{-}6)$$

where G is the cell gravity and $F_{buoyancy}$ is the buoyancy of the cells. Simplifying the above formula produces

$$\nabla E_{ITO}^2 \cdot \vec{z} + \nabla E_{C\text{-}PDMS}^2 \cdot \vec{z} = \frac{2(\rho_{cell} - \rho_{med})g}{3\varepsilon_m \text{Re}[K_{CM}]}, \qquad (2\text{-}7)$$

$$\nabla E^2 \cdot \vec{z} = \frac{2(\rho_{cell} - \rho_{med})g}{3\varepsilon_m \text{Re}[K_{CM}]}. \qquad (2\text{-}8)$$

Substituting Vp-p = 10 V f = 1 MHz, Re[K_{CM}] = –0.06, $\varepsilon_m = 60\varepsilon_0$, $\rho_{cell} = 1.077 \times 10^3$ kg/m³, ρ_{med} = 1.017 × 10³ kg/m³ into Equation (2-8) yields

$$\nabla E^2 \cdot \vec{z} = 1.35 \times 10^{13} \ V^2/m^3.$$

The distribution of the DEP force in X axis and Z axis direction is plotted through simulation analysis, and the distribution curves satisfying Equations (2-3) and (2-7) are shown in Figures 2.29(c) and 2.29(d). The intersection of the two curves in Figure 2.29(e) is the final equilibrium position of the cell. The simulation results show that the cell equilibrium point is located at coordinates (36.3 μm, 59.5 μm).

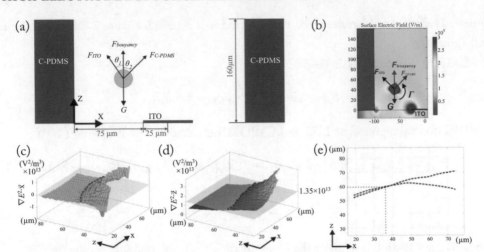

Figure 2.29: Analysis of the equilibrium position of cells in the electrode gap under nDEP force: (a) the force model of cell in the electrode gap; (b) a simulation of cell in the electrode gap; simulation results when the equilibrium conditions are satisfied in the horizontal direction; (c) and the vertical direction (d); and (e) the equilibrium position when the two conditions are satisfied.

Figure 2.30 shows that a HeLa cell is in the middle of the two electrodes under nDEP force (Vp-p = 6 V, f = 600 kHz), and rotates (400°/s) about X-axis under DEP torque. This is consistent with the simulation results.

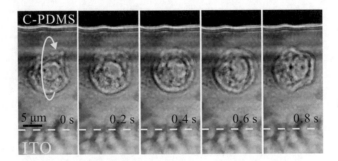

Figure 2.30: One HeLa cell rotated about X-axis between the thick and bottom electrodes under nDEP force.

2.6.4 RELATIONSHIP BETWEEN SPEED AND ELECTRICAL SIGNAL PARAMETERS

DEP torque is proportional to the square of the electric field strength. The greater the signal amplitude, the greater the strength of the electric field, and the greater the rotation speed of the cell. For rotation about Z-axis, when the signal amplitude was below 5 V, cells did not rotate. Rotation occurred when the signal amplitude was greater than 5 V. Figure 2.31(a) shows the rotation speed of a HeLa cell at different amplitudes at frequency $f = 1$ MHz. The relationship between signal amplitude and speed is nonlinear. When the signal amplitude was 5 V, the speed was about ~200°/s. When the signal amplitude increased to 14 V, the speed increased to ~1200°/s which was about 6 times that of 5 V. When the amplitude was greater than 14.5 V, the electrode surfaces were prone to electrolysis and generating bubbles.

For rotation about X/Y-axis, when the signal amplitude was below 5 V, cells did not rotate either. Rotation occurred when signal amplitude was greater than 5 V. Figure 2.31(b) shows the rotation speed of a HeLa cell at different amplitudes at frequency $f = 1$ MHz. The relationship between signal amplitude and speed is also non-linear. When the signal amplitude was 5 V, the speed was about 200°/s. When the amplitude increased to 14 V, the speed increased to ~1000°/s which was 5 times that of 5 V.

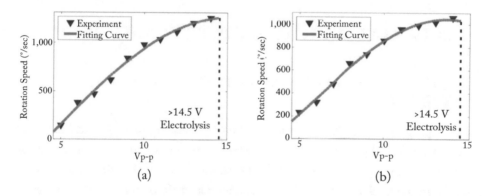

Figure 2.31: Relationship between rotation speed and signal amplitude: (a) in-plane rotation; and (b) out-of-plane rotation.

In addition to signal amplitude, signal frequency is also an important parameter affecting cell speed. At the same signal amplitude, the cell rotation speed is different at different frequencies. Figure 2.32(a) is the in-plane rotation spectrum for a HeLa cell (Vp-p = 6 V). This curve is similar to a bandpass filter, indicating that the cells do not rotate in the low- and high-frequency ranges. Conversely, in the mid-range, the rotation speed of cells increased as the frequency increased first; then the speed decreased as the frequency increased. For out-of-plane rotation, the relationship between speed and frequency is similar to the spectrum of in-plane rotation. Figure 2.32(b) shows the out-of-plane rotation speed spectrum of a HeLa cell (Vp-p = 6 V). The curve resembles the in-plane rotation curve, but with a narrower band. Cells did not rotate in the low-frequency and high-frequency range. In the mid-range, as the frequency increased, the cell rotation speed increased, and after reaching the peak value, the rotation speed decreased as the frequency increased.

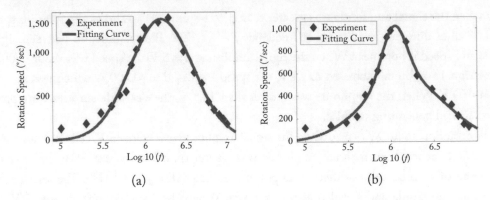

(a)   (b)

Figure 2.32: Relationship between rotation speed and frequency:; (a) in-plane rotation; and (b) out-of-plane rotation.

2.7   CELLULAR ELECTRICAL PROPERTY ANALYSIS

2.7.1   PRINCIPLES OF CELLULAR ELECTRICAL PARAMETER MEASUREMENT

Cellular electrical properties are often used to describe cell viability, growth, and identification of different cell types. Electrical parameters are closely related to the structure and chemical composition of cells, and their physiological functions can be explored by studying the electrical properties. Quantitative analysis of cellular electrical parameters can reflect the dielectric properties of cells and can serve as cell markers. Commonly used cell electrical parameters include the permittivity and conductivity of cell membrane, the permittivity and conductivity of cytoplasm.

Generally, a cell is considered as a single-shell model [168, 169]. The cell is mainly composed of cell membrane and cytoplasm. Assuming that the internal structure of the cytoplasm is uniform, the cell can be equivalent to a single-shell model, as shown in Figure 2.33.

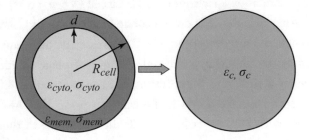

Figure 2.33: Single-shell model and equivalent model.

The equivalent cell complex permittivity is

$$\varepsilon_c^* = \varepsilon_{mem}^* \frac{\left(\frac{R_{cell}}{R_{cell}-d}\right)^3 + 2\left(\frac{\varepsilon_{cyto}^* - \varepsilon_{mem}^*}{\varepsilon_{cyto}^* + 2\varepsilon_{mem}^*}\right)}{\left(\frac{R_{cell}}{R_{cell}-d}\right)^3 - \left(\frac{\varepsilon_{cyto}^* - \varepsilon_{mem}^*}{\varepsilon_{cyto}^* + 2\varepsilon_{mem}^*}\right)} \tag{2-9}$$

where R is the radius of the cell and d is the thickness of cell membrane. ε_{cyto}^* and ε_{mem}^* are the complex permittivity of the cytoplasm and membrane, respectively. $\varepsilon_{cyto}^* = \varepsilon_{cyto} - j\frac{\sigma_{cyto}}{\omega}$, $\varepsilon_{mem}^* = \varepsilon_{mem} - j\frac{\sigma_{mem}}{\omega}$; ε_{cyto} and σ_{cyto} are the permittivity and conductivity of the cytoplasm, ε_{mem} and σ_{mem} are the permittivity and conductivity of the membrane, respectively, and ω are the angular frequencies of the signals.

However, the thickness of cell membrane is generally 8–20 nm, which is difficult to measure. Therefore, the cell area-specific membrane capacitance $C_{mem} = \frac{\varepsilon_{mem}}{d}$ and the cell area-specific membrane conductivity $G_{mem} = \frac{\sigma_{mem}}{d}$ are used instead for cell membrane permittivity and conductivity.

For most mammalian cells, the thickness of the cell membrane is generally much smaller than the cell radius, and its complex permittivity. Equation (2-9) can be equivalent to

$$\varepsilon_c^* = C_{mem}^* \frac{R_{cell} \cdot \varepsilon_{cyto}^*}{R_{cell} \cdot C_{mem}^* + \varepsilon_{cyto}^*}. \tag{2-10}$$

When the cell does rotation motion in solution, Stokes torque can be expressed as

$$\Gamma_f = 8\pi\eta\Omega R_{cell}^3, \tag{2-11}$$

where Ω is the angular velocity of rotation and η is the viscosity of the solution.

When the DEP torque and the Stokes torque are balanced, the cell does uniform rotation

$$\left|\Gamma_{ROT}\right| = \left|\Gamma_f\right|. \tag{2-12}$$

The angular velocity can be expressed as:

$$\Omega = \frac{\varepsilon_m}{2\eta} \text{Im}[K_{CM}] \cdot E^2. \tag{2-13}$$

The typical electro-rotation method of measuring cell dielectric parameters (ε_{mem}^* and ε_{cyto}^*) is to measure cell rotation spectrum and then fit the dielectric parameters of the cell model so that the theoretical electro-rotation spectrum ($\Omega_{theory}(\omega_i)$) is as close as possible to the experimental electro-rotation spectrum $\Omega_{exp}(\omega_i)$:

$$min\sum_i[\Omega_{exp}(\omega_i) - \Omega_{theory}(\omega_i)]^2. \tag{2-14}$$

2.7.2   KEY FACTORS IN ELECTRICAL PROPERTY ANALYSIS

The rotation spectrum is mainly determined by four dielectric parameters (permittivity and conductivity (ε_{mem}, σ_{mem}) of the cell membrane and permittivity and conductivity (ε_{cyto}, σ_{cyto}) of the cytoplasm). The dielectric parameters of different cells are usually different. Without loss of generality, the cell diameter is set to 15 μm, the cell membrane thickness is set to 20 nm according the literature [170], and the rotation spectrum curves related to these four parameters can be plotted according to the parameter settings in Table 2.2.

Table 2.2: Cell dielectric parameter settings

	Conductivity of Cell Membrane σ_{mem} (μS/m)	Permittivity of Cell Membrane ε_{mem}	Conductivity of Cell Cytoplasm σ_{cyto} (S/m)	Permittivity of Cell Cytoplasm ε_{cyto}
Figure 2.34(a)	1–10	$12\varepsilon_0$	0.4	$100\varepsilon_0$
Figure 2.34(b)	10	1–$20\varepsilon_0$	0.4	$100\varepsilon_0$
Figure 2.34(c)	10	$12\varepsilon_0$	0.1–1	$100\varepsilon_0$
Figure 2.34(d)	10	$12\varepsilon_0$	0.4	10–$100\varepsilon_0$

At the whole frequency range, the effect of conductivity of cell membrane on the rotation spectrum is negligible (Figure 2.34(a)). When the frequency is < 100 MHz, the effect of permittivity of cytoplasm on the rotation spectrum is negligible (Figure 2.34(b)). When the frequency is < 100 MHz, permittivity of cell membrane and conductivity of cytoplasm have a significant effect on the rotation spectrum (Figure 2.34(c, d)). This experiment was limited by the signal generator frequency range and the maximum frequency was 20 MHz. Therefore, the method of extracting the dielectric properties of cells can't essentially distinguish conductivity of cell membrane and permittivity of cytoplasm. Therefore, in the experiment, the two parameters were set to $100\varepsilon_0$ and 10^{-6} S/m according to the literature [165].

The electrical parameters of the cell solution also have an effect on the cell rotation spectrum. The permittivity and conductivity of the cell membrane were set to $12\varepsilon_0$ and 10 μS/m, respectively, and the permittivity and conductivity of the cytoplasm were set to $60\varepsilon_0$ and 0.4 S/m, respectively. Figure 2.35 shows the effect of solution permittivity and conductivity on the rotation spectrum.

It can be seen that the permittivity of the solution has significant influence on the rotation spectrum only when the frequency is greater than 100 MHz, and has almost no effect on the rotation spectrum when the frequency is lower than 100 MHz. The change in conductivity of the solution has obviously influence on the rotation spectrum over the whole frequency range, so a lower conductivity (36.5 mS/m) solution was used in the experiment.

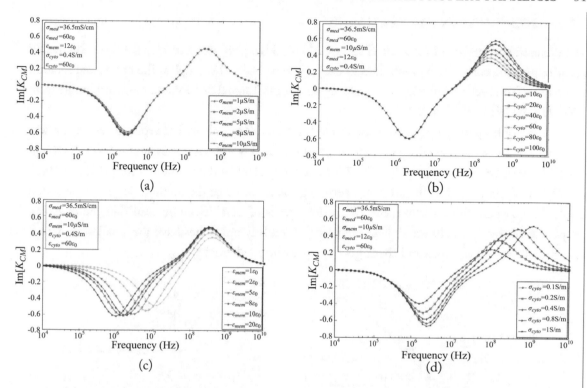

Figure 2.34: Simulation and analysis results of the influence of four electrical parameters on the rotation spectrum (K_{CM} coefficient): (a) effect of cell membrane conductivity; (b) effect of permittivity of cytoplasmic; (c) effect of cell membrane conductivity; and (d) effect of conductivity of cytoplasm.

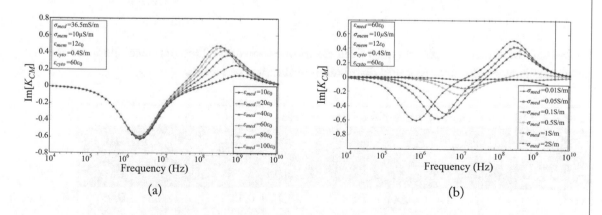

Figure 2.35: Effect of cell solution on the rotation spectrum (K_{CM} coefficient): (a) effect of solution dielectric constant; and (b) effect of solution conductivity.

2.7.3   EXPERIMENTAL ANALYSIS

The dielectric properties of four cells (HeLa, C3H10, B lymphocytes, and HepaRG) were extracted according to the curve fitting method, and each cell type had 8–12 samples. The horizontal rotation spectrum of the cells was obtained by applying an electrical signal of 10 V and frequency range from 100 kHz–10 MHz (Figure 2.36(a)).

Twenty frequency points were measured for each cell sample and the rotation spectrum was fitted with least error. According to the fitting results, the B lymphocyte measurement results (C_{mem} = 10.14 ± 0.08 mF/m²; σ_{cyto} = 0.55 ± 0.07 S/m) are very close to the literature [165] (C_{mem} = 10.33 ± 1.60 mF/m²; σ_{cyto} = 0.41 ± 0.10 S/m), verifying the accuracy of the method. Figure 2.36(b) and Table 2.3 compare the measurements of the four types of cells. It can be seen that the dielectric properties of the four cells are different. Among them, B lymphocytes have the smallest area-specific membrane capacitance and the largest cytoplasmic conductivity.

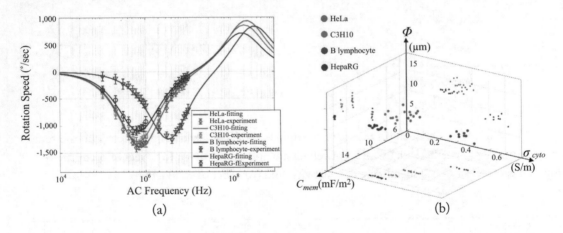

Figure 2.36: Measurement of four types of cell electrical parameters: (a) rotation spectrum measurement and fitting results; and (b) spatial distribution of electrical parameters.

Table 2.3: Electrical characteristics of four types of cells				
Cell Types	Number	Diameter Φ (μm)	Area-specific Membrane Capacitance C_{mem} (mF/m²)	Conductivity of Cytoplasm σ_{cyto} (S/m)
HeLa	10	14.0 ± 1.8	13.11 ± 0.11	0.36 ± 0.05
C3H10	8	14.0 ± 1.5	14.73 ± 0.14	0.31 ± 0.04
B lymphocyte	9	8.2 ± 1.4	10.14 ± 0.08	0.55 ± 0.07
HepaRG	12	12.0 ± 0.5	15.83 ± 0.12	0.26 ± 0.05

2.8   CELL 3D MORPHOLOGY RECONSTRUCTION

2.8.1   PRINCIPLE OF RECONSTRUCTION

Enabled by out-of-plane rotation, a stack of cell contours can be imaged and reconstructed to form 3D morphology. The procedures for 3D reconstruction are as follows.

1. The out-of-plane rotation was recorded and converted into a sequence of frame images.

2. Mark feature points and extract frame images for one revolution.

3. Threshold these frame images into binary ones to identify the cell outline from the background.

4. Remove the background noise and extract the cell outline.

5. Use MATLAB to extract the cell contours of each frame and construct into matrix (2D array).

6. Each image has its own unique angular offset in the 3D model; and the contour points have their 3D coordinates. Then, using MATLAB, transform the contour points to matrix (3D array).

7. Interpolate the 3D array matrix and use the alpha-shape algorithm [171] to reconstruct the 3D morphology of the cell.

Figure 2.37 shows an example of 3D reconstruction of a HeLa cell morphology (241 frames).

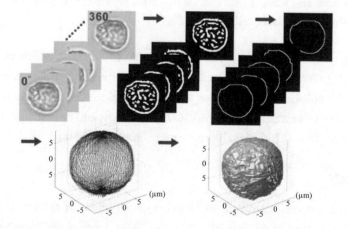

Figure 2.37: 3D reconstruction of cell morphology.

2.8.2   ANALYSIS OF RECONSTRUCTION RESULTS

From the reconstructed 3D cell morphology, several physical features of the cell, such as surface area, volume, roughness, and ellipticity can be calculated. The alpha-shape algorithm builds a 3D model by concatenating point clouds into triangular networks. The calculation of the surface area is obtained by adding the areas of all the triangular meshes. The volume is calculated by using triangular grid projection methods. The roughness is calculated by normalizing the constructed cell model to the standard sphere to determine the corresponding standard deviation. Ellipticity is the ratio of the long axis to the short axis of a 3D model.

The surface area and volume of 8 HeLa cells and 8 B lymphocytes were calculated, as shown in Figure 2.38(a). The average surface area of HeLa cells was 312.16 μm^2, and the average value of B lymphocytes was 180.25 μm^2. The average volume of HeLa cells was 666.86 fL, and the average value of B lymphocytes was 212.03 fL.

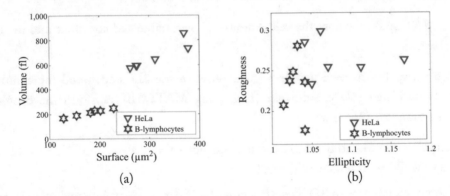

Figure 2.38: (a) Surface area and volume distribution of HeLa cells and B cells; (b) Ellipsoid ratio and surface roughness distribution of HeLa cells and B cells.

The average ellipticity of B lymphocytes was calculated to be 1.029, which was slightly lower than that of HeLa cells (average of 1.084), as shown in Figure 2.38(b). These values indicate that the two cells have similar ellipticities but are not perfectly spherical. On the other hand, the roughness of the cell surface can be calculated from the 3D morphology, and the average roughness of the surface of the B lymphocyte is 0.264 and slightly larger than that of HeLa cells (0.231).

2.9   SUMMARY

This chapter introduced the application of multi-electrode structure of thick-electrode DEP, and presented a single-cell 3D rotation microfluidic device of "Armillary Sphere" type, which is successfully demonstrated for electrical property measurement of single cell and 3D image reconstruction. Control of 3D rotation is achieved by adjusting signal parameters, including rotation direction, speed, and rotation mode. The thick-electrode DEP chip has the advantages of overcoming the limitations of 2D planar electrodes and providing a stable distribution of electric field. In the experiment, the cell area-specific membrane capacitance and cytoplasmic conductivity of cells were extracted by fitting the rotation spectrum. The electrical properties of four types of mammalian cells were measured and analyzed. Based on out-of-plane rotation of cells, the 3D morphology of single cells is reconstructed by image processing, and the geometrical features such as surface area, volume, roughness, and ellipticity are calculated according to the reconstructed 3D morphology. Based on the single-cell 3D rotation chip, it is easy to combine with other techniques to achieve multi-parameter measurements of single cells.

CHAPTER 3

Opto-Electronic Integration of Thick-Electrode DEP Microfluidic Chip

3.1 INTRODUCTION

The multi-electrode structure of thick-electrode DEP overcomes the limitation of electric field attenuation and small working range of the 2D planar electrodes, and enables single-cell 3D rotation. With the development of MEMS technology, the thick-electrode structure can be more easily combined with other technologies to realize multi-function. In order to expand the functions of the multi-electrode structure of thick-electrode DEP, this chapter redesigns the single-cell 3D rotation chip structure, integrates dual-beam optical stretcher into the chip, and realizes multi-parameter measurement of single cells.

Single-cell multi-parameter measurement

Single cells are highly heterogeneous and can be distinguished by cell markers [172, 173]. Cell markers refer to biochemical indicators that can objectively measure and evaluate cells [174, 175]. Examination of cell markers with certain specificity plays an important role in the identification, early diagnosis and prevention of diseases, and therapeutic monitoring. At present, the measurement and characterization of cell markers has gradually become an important hot spot in research. Cell markers are classified into extrinsic biochemical markers [176] and intrinsic markers [177]. The status of cells can be identified by external biochemical markers (e.g., fluorescent pigments, quantum dots, magnetic beads, etc.).

The biophysical markers of cells are intrinsic properties and can be clearly distinguished without label, such as size, shape, density, mechanical, and electrical properties. Intrinsic markers are useful for measuring cell characteristics without biochemical markers. Moreover, similar to the external markers, single intrinsic marker is often insufficient to characterize the cell. To increase specificity, the current research direction is to simultaneously measure multiple intrinsic markers of single cells.

As intrinsic markers of cells, mechanical properties of cells can reflect the cytoskeletal status of cells. The cytoskeleton not only provides the mechanical strength of cells, but also regulates many important cellular functions. The skeleton content and structural changes of a cell can be reflected in the overall mechanical properties of cells. For example, measurements of morphological changes caused by the cytoskeleton can be used to diagnose cancer. On the basis of the measurement of the electrical properties of cells in the previous chapter, the addition of single-cell mechanical property measurement can provide a more comprehensive and in-depth means for specificity analysis of single cells.

3.2   PROGRESS IN SINGLE-CELL MECHANICAL PROPERTY MEASUREMENT

The current main methods for measuring single-cell mechanical properties are as follows.

1. **Magnetic twisting methods.** In 1993, Wang et al. proposed the use of a non-destructive magnetic field to drive a magnetic bead attached to the surface of a cell, exerting a force on the cell, and recording the response of the cell structure under force. The amount of deformation of the cells was calculated via the displacement of the magnetic particles under the action of the magnetic moment, and the influence of the mechanical signals on the structure and function of the cells was quantitatively measured in real time [178]. However, the magnetic properties of the cells can only be tested by magnetic beads, and the results of the tests vary greatly depending on the position of the magnetic beads on the cells [179].

2. **Atomic force microscopy.** Atomic force microscopy relies on the indentation changes produced by the cantilever probe on the cells to measure the mechanical properties of the cells. Atomic force microscopy method is generally suitable for the measurement of adherent cells, but the location of the indentation on a single cell has a large effect on the measurement of mechanical properties [180], as shown in Figure 3.1(a). The measurement point reflects the local mechanical properties of the cell at the 10 nm level due to the probe size, and the throughput of the method is relatively low.

3. **Micropipette aspiration methods.** Mitchson et al. presented the micropipette aspiration method in 1954, as shown in Figure 3.1(b) [181]. The micropipettes having a diameter smaller than the cells are used to aspirate the cells. Under the action of negative pressure, the cells are deformed into the micropipettes, and the mechanical properties of the cells are measured by mapping the relationship between the deformation and the negative pressure. In contrast to atomic force microscopy, micropipette aspiration method is suitable for suspended cells and has been widely used in the measurement of mechanical properties of various white blood cells [182]. However, the results measured by this method are related to the size of the micropipette. The closer the micropipette diameter is to the cell diameter, the better the whole mechanical properties of the cell [183].

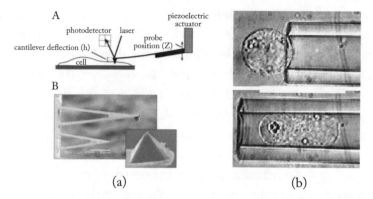

Figure 3.1: (a) Atomic force microscopy [180]; and (b) micropipette aspiration method [181], based on and used with permission from AIP Publishing.

4. **Cell transit analyzers methods**. Cell transit analyzers methods assess cellular mechanical properties by measuring the time the cells take to pass through a microchannel [184, 185]. The advantage of the method is the ability to measure cellular mechanical properties with high throughput, allowing nearly one second to test a cell [186].

To further increase throughput, Rosenbluth et al. proposed a microfluidic chip that simultaneously measures cellular mechanical properties in multiple channels, as shown in Figure 3.2(a). Although the throughput of this method is higher than other measurement methods, this method needs to consider the cell sizes, cell viscosities, and the clogging problem [187].

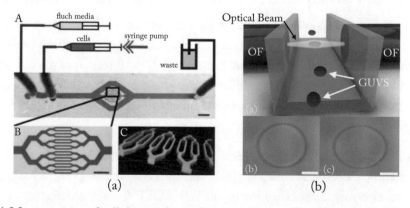

Figure 3.2: (a) Measurement of cellular mechanical properties by cell perforation [187], based on and used with permission from the Royal Society of Chemistry; and (b) dual-beam optical stretching method [188], based on and used with permission from the Royal Society of Chemistry.

5. **Dual-fiber optical stretcher methods.** Figure 3.2(b) is a schematic diagram of the basic structure of a dual-fiber optical stretcher method [188], the principle of which is based on two beams with a Gaussian intensity distribution to form an optical trap. The capture of particles (such as single cells, liposomes, etc.) is achieved by the scattering and refractive power of the beam, and then the stretch deformation is achieved by increasing the optical power [189, 190]. Since the laser beam of the fiber is divergent, the radiation damage to the cell is small. And the optical trap formed by the optical stretcher acts on the whole cell surface, therefore the measured results can directly reflect the mechanical properties of the whole cell. The dual-fiber optical stretcher method is also easy to combine with microfluidic technology to achieve high-throughput single-cell mechanical property measurement [191, 192].

This chapter is based on DEP and optical stretcher techniques to achieve single-cell multi-parameter measurement. Based on the thick-electrode rotation chip of the previous chapter, an optical stretcher was integrated with the thick-electrode structure. The mechanical properties are measured according to the dynamic deformation during the stretching process, and the electrical property measurement is realized by electro-rotation.

3.3 ELECTRO-ROTATION CHIP FUNCTION EXPANSION

3.3.1 ELECTRO-ROTATION CHIP FUNCTION EXPANSION REQUIREMENTS

The thick-electrode DEP techniques have some shortcomings which need to be overcome.

1. **Instability of spatial position when single-cell rotates.** Cells are easily affected by external forces such as flow force. Although they can adapt to position under DEP force, they are susceptible to fluids due to the small DEP force. Once the cell position shifts, the rotation speed will be different because the electric field distribution in the horizontal direction is different, which affects the measurement accuracy of the electrical parameters.

2. **Single-cell manipulation.** Theoretically, mechanical parameters can be obtained by measuring electro-deformation generated by DEP forces [193]. However, the single-cell electro-rotation chip mainly utilizes nDEP force, the deformation is not obvious when the cells are far from electrode, so that accurate mechanical deformation measurement could be very challenging. Based on the simplicity of the structure of the electro-rotation chip, the optical stretcher which is perpendicular to the mi-

crochannel can be easily integrated into the thick electrodes, thereby achieving stable single-cell capture, effective stretching, and mechanical property measurement.

3.3.2   PRINCIPLE OF OPTICAL STRETCHER

The change in light momentum caused by light irradiation on a cell [194, 195] produces axial and gradient forces. The axial force is caused by the collision of photons on the cell, along the direction of propagation of the beam; and the gradient force is caused by the intensity of the light field, and the direction of the light intensity is greatest along the vertical direction of light propagation.

The gradient and axial forces exerted on the cell are related to the laser wavelength and cell size. When the cell radius r is much smaller than the laser wavelength, the gradient and axial forces conform to the Rayleigh model [196], and when cell radius r is much larger than the laser wavelength, the forces conform to the Mie model [197, 198]. For most cells of interest, the cell size corresponds to the Mie model relative to the laser wavelength.

As light enters a cell, the light gains momentum so that the surface gains momentum in the opposite direction. Similarly, the light loses momentum upon leaving the cell so that the opposite surface gains momentum in the direction of the light propagation. The reflection of light on either surface also leads to momentum transfer on both surfaces in the direction of light propagation. For an incident beam, it can be decomposed into multiple beams. Figure 3.3 shows one of the beams with power P_1 incident on the cell.

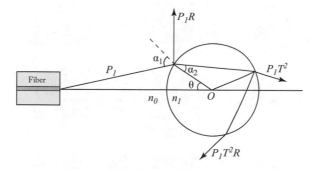

Figure 3.3: Reflection and refraction of the incident light on the cell.

The beam will be reflected and refracted when it enters the cell, and its reflection coefficient R and refractive coefficient T can be expressed as

$$R = \frac{1}{2} \left[\frac{\sin^2 (\alpha_1 - \alpha_2)}{\sin^2 (\alpha_1 + \alpha_2)} + \frac{\tan^2 \alpha_1 - \alpha_2)}{\tan^2 (\alpha_1 + \alpha_2)} \right], \tag{3-1}$$

$$T = 1 - R, \qquad (3\text{-}2)$$

where α_1 is the angle of incidence and α_2 is the angle of refraction.

The Mie model gives the axial force F_{scat} and the gradient force F_{grad} [199]:

$$F_{scat} = \frac{n_0 P_1}{c}\left\{1 + R\cos2\alpha_1 - \frac{T^2\left[\cos(2\alpha_1 - 2\alpha_2)\right] + R\cos2\alpha_1}{1 + R^2 + 2R\cos2\alpha_2}\right\} \qquad (3\text{-}3)$$

$$F_{grad} = \frac{n_0 P_1}{c}\left\{R\sin2\alpha_1 - \frac{T^2\left[\sin(2\alpha_1 - 2\alpha_2)\right] + R\cos2\alpha_1}{1 + R^2 + 2R\cos2\alpha_2}\right\}, \qquad (3\text{-}4)$$

where n_0 is the refractive index of the medium, P_1 is the optical power of the incident beam, c is the speed of light in the vacuum, and n_1 is the refractive index of the cell.

In 2000, Guck et al. first proposed a dual-fiber optical stretcher [195]. Figure 3.4 shows the schematic diagram of the dual-fiber optical stretcher. The dual beams of the oppositely propagated laser beam can capture and stretch single cells. According to the principle of conservation of light momentum, when the laser is incident on the cell, the front and back surfaces of the cell are subjected to an outward force, and the direction is perpendicular to the cell surface. When the cells are in the middle of the optical fibers, the force equilibrium can be achieved in the axial direction and the normal direction to achieve single-cell capture, and then the single cell can be stretched and deformed by increasing the optical power.

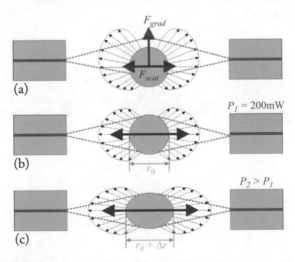

Figure 3.4: The working principle of optical stretcher: (a) the cell are subjected to axial and gradient forces when it offsets from the fiber axis; and (b) the cell is trapped; (c) the cell is stretched when optical power is increased.

The dual-fiber optical stretcher has the following advantages: (1) the beam is unfocused, so the power on the cell is not high, and the damage to the cell is negligible; (2) the capture and stretching of cells can be easily achieved; and (3) the dual-fiber optical stretcher is low in cost, and the operation is simple and easy to combine with microfluidic technologies. However, in order to achieve stable capture, the optical stretcher must use single-mode fiber, while the single-mode fiber has a core diameter of only 9 μm, which requires accurate alignment of the fibers. This imposes some difficulty in chip fabrication.

3.3.3   STEP-STRESS ANALYSIS OF CELL MECHANICAL PROPERTIES

The mechanical properties of the cells can be measured by step-stress tests on single cells [200]. Δr is the deformation of the cells along the optical axis, and r is the original diameter of the cell. Figure 3.5 is a schematic diagram of the step-stress response curve of a cell.

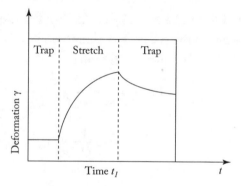

Figure 3.5: Step stress response curve of cells.

Cell mechanical properties can be fitted by the following formula describing the response curve:

$$\gamma(t) = \sigma_0 \left(\frac{b_1}{a_1} - \frac{a_2}{a_1^2} \right) \left(1 - \exp\left(-\frac{a_1}{a_2} t \right) \right) + \frac{\sigma_0}{a_1} t, \qquad (3\text{-}5)$$

where a_1, a_2, and b_1 are parameters of the fitted curve. σ_0 is the maximum stress applied to the cell along the beam:

$$\sigma_0 = \frac{n_{med} I_0}{c} (2 - R_{ref} + R_{ref}^2)\left(\frac{n_{cell}}{n_{med}} - 1\right), \qquad (3\text{-}6)$$

where c is the velocity of light in vacuum, n_{med} is the refractive index of the medium (normally $n_{med} \geq 1.335$), n_{cell} is the refractive index of the cell, and R is the amount of light reflected at the interface between the medium and the cell, which can be expressed as

$$R_{ref} = \left(\frac{n_{cell} - n_{med}}{n_{cell} + n_{med}}\right)^2. \tag{3-7}$$

I_0 is the laser intensity along the laser axis at the position of the cell and can be calculated using the following equation:

$$I_0 = \frac{2P}{\omega^2 \pi}, \tag{3-8}$$

where P is the total power of the laser beams and ω is the radius of the laser beams at the position of the cell.

When the laser is turned off, the shape of the cell will gradually recover. Upon the laser power retraction, the cell relaxes dynamically back to equilibrium status and the strain can be fitted by

$$\gamma(t) = \sigma_0 \left(\frac{b_1 - a_2}{a_1}\frac{}{a_1^2}\right)\left(1 - \exp\left(-\frac{a_1}{a_2}t\right)\right)\exp\left(-\frac{a_1}{a_2}(t - t_1)\right) + \frac{\sigma_0}{a_1}t. \tag{3-9}$$

The parameters a_1, a_2, and b_1 can be fitted by combining Equations (3-5) and (3-9). Then the mechanical property parameters of the cell such as shear modulus G, steady-state viscosity η or relaxation time τ can be calculated by the following formula:

$$G = \frac{1}{2(1+\mu)}\frac{a_1^2}{a_1 b_1 - a_2}, \tag{3-10}$$

$$\eta = \frac{1}{2(1+\mu)}a_1, \tag{3-11}$$

$$\tau = b_1, \tag{3-12}$$

where μ is the Poisson's ratio (normally, $\mu \approx 0.45$–0.50) [200].

3.4   CHIP DESIGN AND FABRICATION

The chip consists of a microchannel with thick electrodes, two optical fibers, and a substrate. The structure is shown in Figure 3.6. The optical fiber (125 μm in diameter) is placed perpendicular to the microchannel. To ensure single-cell capture and stretch, the two fibers need to be aligned well in three dimensions.

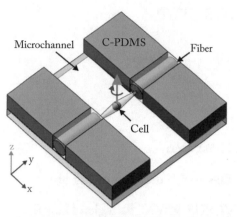

Figure 3.6: The structure of the electro-rotation and optical stretch chip.

The fabrication procedures of the chip are shown in Figure 3.7(a): (1) fabricate the SU-8 mold; (2) pour C-PDMS on the SU-8 mold; (3) scrap off the excess C-PDMS with a blade, and cure at 80°C for 2 h; (4) pour PDMS on the cured C-PDMS, remove bubble and cure again; (5) remove the PDMS-CPDMS from the SU-8 mold; (6) punch the holes; (7) align the PDMS-CPDMS with the patterned ITO glass; (8) lead electrodes through the wire; and (9) insert two single-mode fibers. In order to ensure the sealing of the microchannel, glue is used for sealing outside the chip. Figure 3.7(b) is the photo of chip.

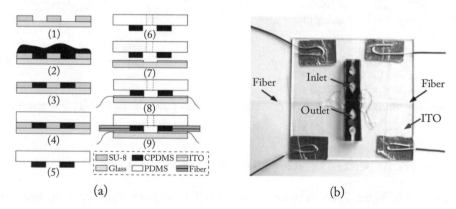

Figure 3.7: Chip processing: (a) schematic diagram of the chip processing flow; and (b) chip processing physical map.

3.5   EXPERIMENTAL SETUP

3.5.1   EXPERIMENTAL INSTRUMENTS

Equipment about optical stretcher

1. Single-mode fiber (HI1060, Corning)

   HI1060 single-mode fiber with a diameter of 125 μm.

2. 980 nm pump laser (VENUS-980-M, Shanghai Haoyu)

   A pump laser with a wavelength of 980 nm is used, as shown in Figure 3.8(a), and the power adjustment range is from 0–850 mW. The 980-nm laser wavelength is in the near-infrared wavelength region, and the absorption rate of the light source in water is relatively low, so the damage to cells is negligible.

(a)   (b)

Figure 3.8: Dual-fiber stretcher optical path device: (a) 980-nm pump laser; and (b) 980-nm optocoupler.

1. 980-nm optical coupler (VENUS-980-M, Shanghai Haoyu)

   As shown in Figure 3.8(b), the optocoupler splits the laser from the laser into two beams, with a split ratio of 50/50, and the two beams are connected to two single-mode fibers.

2. Optical isolator (OI-980, Shanghai Haoyu)

   The optical isolator is used to prevent laser propagating from one of the opposite fibers from returning into the laser through another optical fiber to damage the laser. The optical isolator, as shown in Figure 3.9(a), is connected between the output of the laser and the optical wavelength division multiplexer, which has a small forward transmission loss to the 980-nm laser and a large reverse transmission loss.

3. Filter mirror (YZ-532LGP-Y68, Yizheng Laser)

The 980-nm laser easily causes the pixels on the CCD camera to be overexposed and appear as white spots. In order to facilitate the observation of the response of the cells, a 980-nm filter mirror is placed in front of the CCD camera. Figure 3.9(b) is photo of a 980-nm filter mirror.

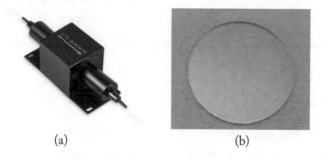

(a)   (b)

Figure 3.9: (a) 980-nm laser isolators; and (b) 980-nm filter mirror.

The working diagram and setup of the experiment are shown in Figure 3.10. The chip is placed under an inverted microscope (Nikon Ti-U) and the four electrodes of the chip are connected to a four-channel signal generator (TGA-12104, TTi). The dual fibers are passed through an optocoupler that is connected to a 980-nm pump laser. The inlet and outlet are led out through a Teflon tube. In this experiment, gravity is used to drive the flow, and the container of cell solution is placed on the 3D manipulator (MP285, Sutter) which can move up and down to change the height of the container, and the movement precision can reach 40 nm. Experimental image and video capture are performed using an industrial camera (acA 640-120 gm, Basler).

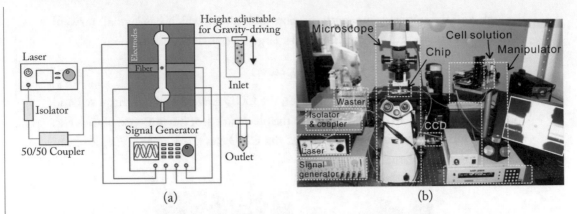

Figure 3.10: Experimental setup: (a) schematic diagram; and (b) photo of the experimental platform.

3.5.2   EXPERIMENTAL STEPS

Figure 3.11 shows the working procedures of the experiment.

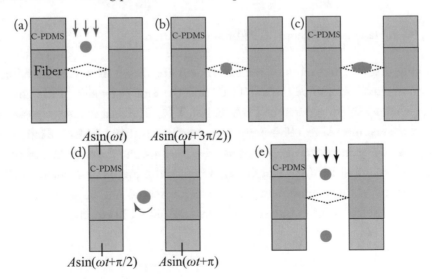

Figure 3.11: The working procedures of the experiment.

a. **Cell loading:** The height of the cell solution container is controlled by a 3D manipulator, and the flow rate is controlled by adjusting the height of the cell solution container.

b. **Cell Capture:** When the cell approaches the fiber capture zone, it is first pulled to the center of the light intensity by gradient force, until an equilibrium position

c. **Cell stretch:** After the cell is captured, the axial force acting on the cell is increased by increasing the laser power, causing the cell to axially stretch.

d. **Cell rotation:** After the cells have completed the tensile deformation and restored to the original state. Proper electrical signals are applied to the four thick electrodes to cause the cell to rotate horizontally. The electrical properties of the cell are measured by fitting analysis of the rotation spectrum.

e. **Cell recovery:** After the rotation is completed, the height of the cell solution container is lifted to flush the cell away from the working area to recover.

3.6   SINGLE-CELL MANIPULATION AND MULTI-PARAMETER ANALYSIS EXPERIMENTS

The cell sample preparation in the experiment was the same as that in Section 2.6.1. Five types of cells are selected in this experiment, HeLa cells, A549 cells, HepaRG cells, MCF7 cells, and MCF10A cells.

3.6.1   EXPERIMENTAL DEMONSTRATION OF FILTER MIRROR

In experiment, a 980-nm filter was placed in front of the CCD camera to eliminate the influence of CCD overexposure. Figure 3.12 shows the effect of the filter on dual-beam laser capture of a MCF10A cell. If there was no filter, the laser scattered on the cell, and the image on the CCD is flaring, and the cell contour was hardly observable (Figure 3.12(a)). Figure 3.12(b) shows the image with the filter, effectively eliminating the effects of overexposure caused by scattered light.

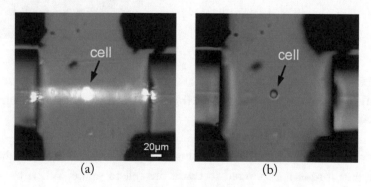

(a)   (b)

Figure 3.12: Dual beam capture of one MCF10A cells: (a) without filter and (b) with filter.

3.6.2   CELL MOTIONS WHEN THE FIBERS ARE MISALIGNED

When the two fibers are not perfectly aligned, the force acting on the cells is asymmetric, and stable cell capture can't be achieved. According to the deviation of the center of the two fibers, the cells can be captured or rotated. When the deviation L of the center position of the fiber is larger than the cell radius r_{cell}, as shown in Figure 3.13(a), the cell is pushed onto the other fiber end by one of the laser beams. Figure 3.13(b) shows a HeLa cell contact on the fiber end (laser power 200 mW).

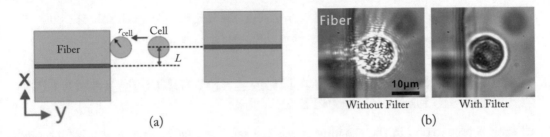

(a)   (b)

Figure 3.13: Cell cannoy be captured when the fiber alignment deviation is greater than the cell radius: (a) schematic diagram of fiber alignment deviation; and (b) one HeLa cell is captured on the fiber end face.

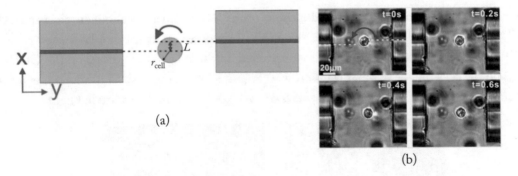

(a)   (b)

Figure 3.14: Dual fibers that are misaligned in the horizontal direction cause horizontal rotation of the cells: (a) schematic diagram of the principle; and (b) one A549 cell was rotated counterclockwise under the action of double fibers.

When the deviation L of the dual fibers is smaller than the cell radius r_{cell}, the force on the cell is not collinear, and torque is generated to cause the cell to rotate. When the misalignment occurs in the horizontal direction, the torque is in the horizontal direction, causing the cells to rotate horizontally. Figure 3.14 shows that one A549 cell rotating in the horizontal direction of the misaligned dual fibers (laser power 200 mW).

When the misalignment occurs in the vertical direction, the torque is in the vertical direction, causing the cells to do out-of-plane rotation. Figure 3.15 shows a MCF7 cell in the vertical direction misaligned by the dual fibers (laser power 200 mW) doing out-of-plane rotation.

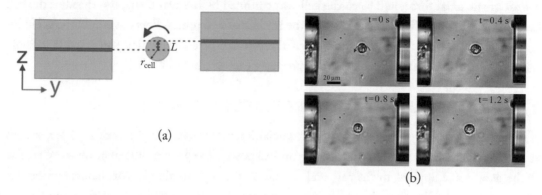

Figure 3.15: Dual fibers that are misaligned in the vertical direction causes the cells to do out-of-plane rotation: (a) schematic diagram of the principle: and (b) one MCF7 cell did out-of-plane rotation under the action of dual fibers (laser power 200 mW).

3.6.3   SINGLE-CELL DUAL-FIBER CAPTURE EXPERIMENT

When a cell flows through dual fibers, the gradient force generated by the laser on the cell pulls the cell toward the position where the intensity distribution is greatest, that is the center of the axis of the fiber. When the cell is off center of the axis, the gradient force pulls the cell back to the center of the axis. The axial force generated on the cell pushes the cell away from the fiber end. Because the two fibers are relatively distributed, a force balance is formed between the two fibers. A single cell can be stably captured by the combination of gradient force and axial force.

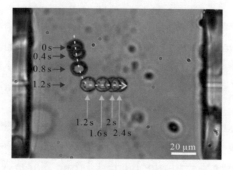

Figure 3.16: The process of capturing one HeLa cell by dual fibers.

Figure 3.16 shows the optical trap process of a HeLa cell. Each power of optical fiber is 100 mW. When the cell was close to the position of the optical axis, the cell was first accelerated to the optical axis with the highest light intensity by the axial force. Then the cell was pushed away under the action of the axial force, and since the cell was confined by the two fibers, the speed of the cell gradually decreased. Finally, it was trapped in the middle of two optical fibers at equilibrium. Under the action of the double fibers, the position of the cell can be stable, which can provide a stable spatial position for mechanical and electrical property measurement of cells.

3.6.4   SINGLE-CELL OPTICAL STRETCH EXPERIMENT

Since the strength of the gradient force and the axial force are related to the optical power, stretch of the single cell can be achieved by increasing optical power. The greater optical power applied, the larger the axial force applied to the cell, and the larger the deformation is. Mechanical properties of different types of cells are different, and the corresponding deformations are different under the same optical power.

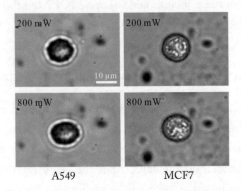

Figure 3.17: Comparison of stretch deformation of A549 and MCF7 cells at the same power.

Figure 3.17 shows the results of stretch deformation of two cells (A549 and MCF7). It can be found that the two cells can maintain a stable spatial position at a total power of 200 mW. When increasing the optical power to 800 mW, both cells have significant stretch deformations. The maximum deformation of A549 cells reached 14%, while the deformation of MCF7 was smaller, only 8%.

To measure the relationship between cell deformation and optical power, A549 cells were used as an example to measure the deformations of cells at different optical powers, as shown in Figure 3.18(a). The optical power is gradually increased from 200 mW to 800 mW, and the cell deformation is gradually increased. The cell deformation and the optical power have a good linear relationship. The stretch deformation of the five types of cells (10 per cell sample) at 800 mW were measured, as shown in Figure 3.18(b). HepaRG cells have the smallest deformation (about 5%), while A549 has the largest deformation (about 15%), and MCF7 has about twice the deformation of normal breast cell MCF10A.

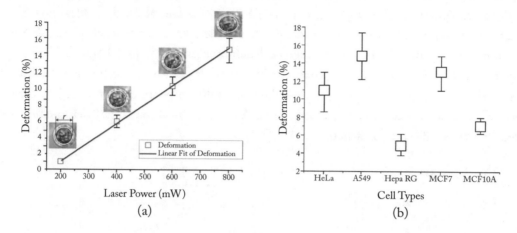

Figure 3.18: (a) The deformations of A549 cells at different optical powers, and (b) the deformations of five types of cells at 800 mW.

In addition to capturing individual cells, dual fibers are capable of simultaneously capture and stretch multiple cells. When multiple cells are located in the center of a dual fibers, the cells form a cells chain under the optical force, and increasing the optical power can simultaneously stretch the cell chain. Figure 3.19 shows the simultaneous capture of two MCF10A cells (200 mW) and stretch (800 mW) with deformation of approximately 4.5%.

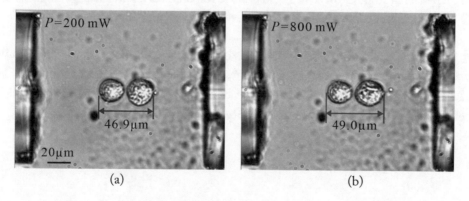

Figure 3.19: Optical stretcher captured and stretched two cells: (a) cell captured; and (b) cell stretched.

3.6.5   SINGLE-CELL MECHANICAL PROPERTY MEASUREMENT

From the theoretical analysis of Section 3.3.3, the mechanical properties of the cell can be calculated by measuring the step-stress response of the cells. The cells were captured using 200 mW of power in the experiment. After applying a laser of 800 mW power of 5s, the optical power was reduced to 200 mW, and the stretched and relaxed state of the cells during this period was observed, and the time-varying curve of the corresponding cell deformation was plotted. Figure 3.20(a) shows the response curves for 10 A549 cells, where the blue curve is the average of the deformation responses of all 10 samples. Figure 3.20(b) plots the mean value response curve for each of the five types of cells measured (10 samples per cell type). A549 cells have the largest deformation and HepaRG has the smallest deformation.

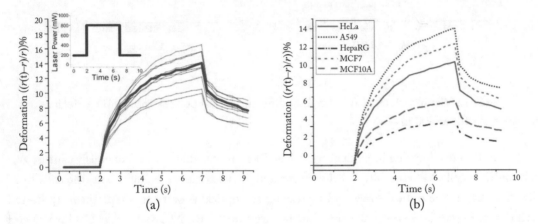

Figure 3.20: Step-stretch response of cells: (a) step-stress response curves and average response curves for 10 A549 cell samples; and (b) average response curves for five types of cells.

By fitting the step-stress response curves of the five types of cells, the mechanical property results were calculated, as shown in Table 3.1. The results showed that the shear modulus of HeLa, A549, and MCF7 cancer cells was lower than that of HepaRG and MCF10A normal cells. And HepaRG cells have the highest steady-state viscosity and relaxation time.

Table 3.1: Test results of five types of cellular mechanical properties

Cell Types	Shear Modulus G (Pa)	Steady-State Viscosity η (Pa·s)	Relaxation Time τ (s)
HeLa	19.10 ± 0.81	62.13 ± 4.30	3.50 ± 0.383
A549	14.63 ± 0.65	47.93 ± 3.37	3.53 ± 0.396
HepaRG	44.99 ± 1.97	156.37 ± 11.27	3.72 ± 0.422
MCF7	15.98 ± 0.69	52.07 ± 3.67	3.51 ± 0.392
MCF10A	30.67 ± 1.30	100.47 ± 6.83	3.53 ± 0.381

3.6.6   SINGLE-CELL ELECTRO-ROTATION

A single cell can be held in stable space for a long time after being captured by dual fibers. In order to avoid the influence of the laser on the electro-rotation, the laser needs to be turned off when the electro-rotation signal is applied. If the cell sinks down, the laser can be re-opened to drag the cell back to the capture position. In-plane rotation of the cell is achieved by applying electrical signals to the four thick electrodes. Figure 3.21 shows the corresponding electric field simulation result. The electric field in the electrode chamber rotates clockwise (Vp-p = 10 V, f = 1 MHz) in one signal period, indicating that fibers don't affect the distribution of electric field in the chamber.

Figure 3.22(a) shows clockwise rotation of HeLa cells under an electric field (Vp-p = 6V, f = 600 kHz) at a speed of approximately 60°/s. The direction of cell rotation can be changed by switching the phase of the electrical signals. Figure 3.22(b) shows that the HepaRG cells rotate counterclockwise under the electric field (Vp-p = 6V, f = 600 kHz) and the rotation speed is about 45°/s.

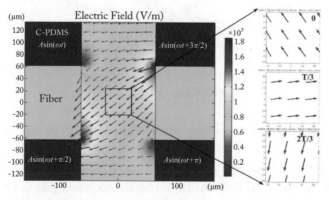

Figure 3.21: Simulation diagram of electro-rotation.

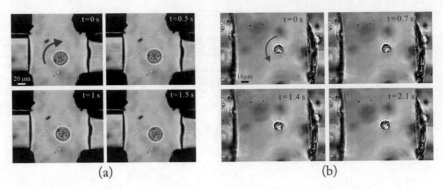

Figure 3.22: Cell electro-rotation: (a) one HeLa cell did clockwise rotation (Vp-p = 6V, f = 600 kHz) and (b) one HepaRG cell did clockwise rotation (Vp-p = 6V, f = 600 kHz).

It can be seen from Equation (1-10) that the DEP torque acting on the cell is proportional to the electric field strength. The frequency f was set as 1 MHz to derive the relationship between the rotation speed and applied voltage for the five types of cells, as shown in Figure 3.23. It can be seen that the larger the voltage amplitude, the greater the rotation speed of the cells. At the same voltage amplitude, there are also differences in the rotation speeds for different types of cells. In particular, HeLa cells rotate fastest and A549 lowest.

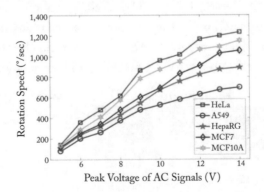

Figure 3.23: Electro-rotational spectra and fitting results of five types of cell measurements.

When measuring the rotation spectrum, the signal amplitude was set to Vp-p = 10 V, the frequency range was chosen from 100 kHz to 10 MHz. The electrical parameters of the cells were calculated by fitting the rotation spectrum (shown by the solid line in Fig. 3.24(a)). Figure 3.24(b) and Table 3.2 show the electrical parameter results of the five types of cells. A549 cells have the largest area-specific membrane capacitance and the smallest cytoplasmic conductivity relative to the other four types of cells. MCF10A cells have the smallest area-specific membrane capacitance and the maximum cytoplasmic conductivity; the average cell area-specific membrane capacitance of MCF7 is about 1.5 times that of MCF10A.

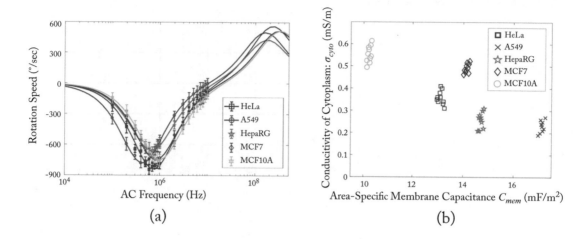

Figure 3.24: (a) Electro-rotation spectra and fitting results of five types of cells; and (b) distribution of electrical parameters corresponding to five types of cells.

Table 3.2: Electrical parameters of five types of cells

Cell Types	Area-Specific Membrane Capacitance C_{mem} (mF/m^2)	Cytoplasmic Conductivity σ_{cyto} (S/m)
HeLa	13.23 ± 0.32	0.35 ± 0.06
A549	17.12 ± 0.14	0.23 ± 0.04
HepaRG	14.71 ± 0.15	0.23 ± 0.07
MCF7	14.15 ± 0.09	0.50 ± 0.03
MCF10A	10.16 ± 0.07	0.55 ± 0.02

3.7   SUMMARY

This chapter is the extension of the function of multi-electrode chips for thick-electrode DEP. Based on the thick-electrode DEP multi-electrode 3D rotation chip, the thick-electrode DEP multi-electrode structure integrated with the optical stretcher is presented to achieve multiple physical parameter measurements of single cells. The single cell was first captured using an optical stretcher and the mechanical properties of the cell were measured by step-stress experiments. The optical stretcher provides a stable spatial position via single-cell capture for single-cell rotation, which not only ensures the stability of cell electro-rotation but also improves the accuracy of measurement of electrical parameters. The experiment results show that the five types of cells have different mechanical and electrical properties. The use of this opto-electronic integration chip expands the application of thick-electrode DEP.

CHAPTER 4

Summary and Outlook

4.1 MAIN WORK

Based on the basic structure of thick-electrode DEP, the multi-electrode composition of thick-electrode DEP was studied. A single-cell 3D rotation chip of "Armillary Sphere" was proposed and fabricated. The chip design combined with single-cell capture V-shape structure. The trap structure and the 3D rotation thick-electrode structure solve the problem of single-cell loading difficulty. Single-cell 3D rotation is achieved by applying different electrical signal configurations on the electrodes. Based on single-cell 3D rotation manipulation, single-cell electrical parameter measurement and 3D morphology reconstruction are realized. The reconstructed 3D morphology can be further used to extract the physical features of single cells, such as surface area, volume, and ellipsoid rate, providing a new operating platform for single-cell analysis.

To realize the opto-electronic integration application of multi-electrode structure with thick-electrode DEP, a single-cell optical stretcher and electro-rotation chip was presented. On the basis of the single-cell electro-rotation chip, a dual-fiber optical stretcher was integrated for single cell capture and mechanical property measurement. The high-efficiency single-cell capture ensures the stable spatial position of the cells, and combined with electro-rotation, can perform single-cell mechanical and electrical characteristic parameters measurement on the same chip. Five types of cells were measured in the experiment. The combination of mechanical and electrical properties confirmed the specificity of single cells, further broadening the application of thick-electrode DEP in microfluidics.

4.2 MAJOR INNOVATIONS

1. Proposed and prepared a single-cell 3D rotation chip with structure resembling "Armillary Sphere," which overcomes the limitation of electric field of planar electrodes, realizes single-cell 3D rotation, electrical parameter measurement, and 3D image reconstruction.

2. Presented a single-cell optical stretcher-electro-rotation chip with dual fibers embedded into thick electrodes to simultaneously measure single-cell mechanical and electrical properties, providing a versatile operating platform for single-cell studies.

4.3   FUTURE PROSPECTS

In this work, C-PDMS was used as the thick electrode material. Although the thick electrodes are easy to process, the conductivity is not as high as metals. With the continuous development of MEMS technology, more materials and sophisticated processing methods can be used to make thick electrodes, and the application of thick-electrode DEP in the field of biological microfluidics can be further explored.

The multi-electrode structure of the thick-electrode DEP-single-cell 3D rotational chip also has the problem that the chip is non-open structure, which makes trouble with single-cell recovery. In addition, for 3D cell imaging, a more advanced algorithm can be developed to retrieve the intracellular information (e.g. nuclear-to-cytoplasmic ratio and organelle structure) like what is done for CT or MRI scanning, not limited to the morphology alone. At the same time, the cell activity and functionality after the operation have not been studied in the current experiment.

The opto-electronic integration chip can be subsequently used for the separation of cell samples due to its multi-parameter measurement function. However, the optical fiber core is small such that the capture range is small. The problem greatly limits the efficiency of capture and stretch. In future work, it can be combined with other technical means to achieve the position control of the cells in the microchannel, and improve the efficiency of fiber capture and stretch.

References

[1]   Thorsen, T. Microfluidic Large-Scale Integration. *Science*. 2002; 298(5593):580–584. DOI: 10.1126/science.1076996. 1, 2

[2]   Lee, B. S., Lee, J. N., Park, J. M., Lee, J. G., Kim, S., Cho, Y. K., and Ko, C. A fully automated immunoassay from whole blood on a disc. *Lab on a Chip*. 2009; 9(11):1548–1555. DOI: 10.1039/b820321k. 1

[3]   Li, R., Jia, F., Zhang, W., Shi, F., Fang, Z., Zhao, H., Hu, Z., and Wei, Z. Device for whole genome sequencing single circulating tumor cells from whole blood. *Lab on a Chip*. 2019; 19(19): 3168–3178. DOI: 10.1039/c9lc00473d. 1

[4]   Manz, A., Graber, and N., Widmer, H. M. Miniaturized total chemical analysis systems: A novel concept for chemical sensing. *Sensors and Actuators B: Chemical*. 1990; 1(1-6):244–248. DOI: 10.1016/0925-4005(90)80209-I. 1

[5]   Harrison, D. J., Manz, A , Fan, Z., Luedi, H., and Widmer, H. M. Capillary electrophoresis and sample injection systems integrated on a planar glass chip. *Analytical Chemistry*. 2002; 64(17):1926–1932. DOI: 10.1021/ac00041a030. 1

[6]   Harrison, D. J., Fluri, K., Seiler, K., Fan, Z., Effenhauser, C. S., and Manz, A. Micromachining a miniaturized capillary electrophoresis-based chemical analysis system on a chip. *Science*. 1993; 261(5123):895–897. DOI: 10.1126/science.261.5123.895. 2

[7]   Woolley, A. T. and Mathies, R. A. Ultra-high-speed DNA sequencing using capillary electrophoresis chips. *Analytical Chemistry*. 2002; 67(20):3676–3680. DOI: 10.1021/ac00116a010. 2

[8]   Woolley, A. T., Hadley, D., Landre, P., deMello, A. J., Mathies, R. A., and Northrup, M. A. Functional integration of PCR amplification and capillary electrophoresis in a microfabricated DNA analysis device. *Analytical Chemistry*. 1996; 68(23):4081–4086. DOI: 10.1021/ac960718q. 2

[9]   Woolley, A. T., Lao, K., Glazer, A. N., and Mathies, R. A. Capillary electrophoresis chips with integrated electrochemical detection. *Analytical Chemistry*. 1998; 70(4):684–688. DOI: 10.1021/ac971135z. 2

[10]   Burns, M. A. An integrated nanoliter DNA analysis device. *Science*. 1998; 282(5388):484–487. DOI: 10.1126/science.282.5388.484. 2

[11]   Weigl, B. H., Bardell, R. L., and Cabrera, C. R. Lab-on-a-chip for drug development. *Advanced Drug Delivery Reviews*. 2003; 55(3):349–377. DOI: 10.1016/S0169-409X(02)00223-5. 2

[12]   Erickson, D. Towards numerical prototyping of labs-on-chip: modeling for integrated microfluidic devices. *Microfluidics and Nanofluidics*. 2005; 1(4):301–318. DOI: 10.1007/s10404-005-0041-z. 2

[13]   Whitesides, G. M. The origins and the future of microfluidics. *Nature*. 2006; 442(7101):368–373. DOI: 10.1038/nature05058. 2

[14]   Bhatia, S. N. and Ingber, D. E. Microfluidic organs-on-chips. *Nature Biotechnology*. 2014; 32(8):760–772. DOI: 10.1038/nbt.2989. 2

[15]   Mark, D., Haeberle, S., Roth, G., von Stetten, F., and Zengerle, R. Microfluidic lab-on-a-chip platforms: requirements, characteristics and applications. *Chemical Society Reviews*. 2010; 39(3):1153. DOI: 10.1039/b820557b. 2

[16]   Pan, P., Wang, W., Ru, C., Sun, Y., and Liu, X. MEMS-based platforms for mechanical manipulation and characterization of cells, *Journal of Micromechanics and Microengineering*, 2017, 27, 123003. DOI: 10.1088/1361-6439/aa8f1d. 3

[17]   Qu, B. Y., Wu, Z. Y., Fang, F., Bai, Z. M., Yang, D. Z., and Xu, S. K. A glass microfluidic chip for continuous blood cell sorting by a magnetic gradient without labeling. *Analytical and Bioanalytical Chemistry*. 2008; 392(7-8):1317–1324. DOI: 10.1007/s00216-008-2382-4. 3

[18]   Sayah, A., Thivolle, P-A., Parashar, V. K., and Gijs, M. A. M. Fabrication of microfluidic mixers with varying topography in glass using the powder-blasting process. *Journal of Micromechanics and Microengineering*. 2009; 19(8):085024. DOI: 10.1088/0960-1317/19/8/085024. 3

[19]   Ino, K., Ishida, A., Inoue, K. Y., Suzuki, M., Koide, M., Yasukawa, T., Shiku, H., and Matsue, T. Electrorotation chip consisting of 3D interdigitated array electrodes. *Sensors and Actuators B: Chemical*. 2011; 153(2):468–473. DOI: 10.1016/j.snb.2010.11.012. 3

[20]   Dochow, S., Beleites, C., Henkel, T., Mayer, G., Albert, J., Clement, J., Krafft, C., and Popp, J. Quartz microfluidic chip for tumour cell identification by Raman spectroscopy in combination with optical traps. *Analytical and Bioanalytical Chemistry*. 2013; 405(8):2743–2746. DOI: 10.1007/s00216-013-6726-3. 3

[21]   Sabounchi, P., Morales, A. M., Ponce, P., Lee, L. P., Simmons, B. A., and Davalos, R. V. Sample concentration and impedance detection on a microfluidic polymer chip. *Biomedical Microdevices*. 2008; 10(5):661–670. DOI: 10.1007/s10544-008-9177-4. 3

[22]   Meagher, R. J., Coyne, J. A., Hestekin, C. N., Chiesl, T. N., Haynes, R. D., Won, J-I., and Barron, A. E. Multiplexed p53 mutation detection by free-Solution conjugate microchannel electrophoresis with polyamide drag-tags. *Analytical Chemistry*. 2007; 79(5):1848–1854. DOI: 10.1021/ac061903z. 3

[23]   Ogończyk, D., Węgrzyn, J., Jankowski, P., Dąbrowski, B., and Garstecki, P. Bonding of microfluidic devices fabricated in polycarbonate. *Lab on a Chip*. 2010; 10(10):1324–1327. DOI: 10.1039/b924439e. 3

[24]   Zhang, H. and Sun, Y. Optofluidic droplet dye laser generated by microfluidic nozzles. *Optics Express*. 2018; 26(9):11284. DOI: 10.1364/OE.26.011284. 3

[25]   Li, J., Chen, D., and Chen, G. Low-Temperature thermal bonding of PMMA microfluidic chips. *Analytical Letters*. 2005; 38(7):1127–1136. DOI: 10.1081/AL-200057209. 3

[26]   Ogilvie, I. R. G., Sieben, V. J., Cortese, B,. Mowlem, M. C., and Morgan, H. Chemically resistant microfluidic valves from Viton® membranes bonded to COC and PMMA. *Lab on a Chip*. 2011; 11(14):2455–2459. DOI: 10.1039/c1lc20069k. 3

[27]   Berdichevsky, Y., Khandurina, J., Guttman, A., and Lo, Y. H. UV/ozone modification of poly(dimethylsiloxane) microfluidic channels. *Sensors and Actuators B: Chemical*. 2004; 97(2-3):402–408. DOI: 10.1016/j.snb.2003.09.022. 3

[28]   Kim, B-Y., Hong, L-Y., Chung, Y-M., Kim, D-P., and Lee, C-S. Solvent-resistant PDMS microfluidic devices with hybrid inorganic/organic polymer coatings. *Advanced Functional Materials*. 2009; 19(23):3796–3803. DOI: 10.1002/adfm.200901024. 3

[29]   Feng, Y., Huang, L., Zhao, P., Liang, F., and Wang, W. A microfluidic device integrating impedance flow cytometry and electric impedance spectroscopy for high-efficiency single-cell electrical property measurement, *Analytical Chemistry*, 2019, 91(23), 15204–15212, DOI: 10.1021/acs.analchem.9b04083. 3

[30]   Liu, H., and Crooks, R. M. 3D Paper microfluidic devices assembled using the principles of origami. *Journal of the American Chemical Society*. 2011; 133(44):17564–17566. DOI: 10.1021/ja2071779. 3

[31]   Dou, M., Sanjay, S. T., Benhabib, M., Xu, F., and Li, X. Low-cost bioanalysis on paper-based and its hybrid microfluidic platforms. *Talanta*. 2015; 145:43–54. DOI: 10.1016/j.talanta.2015.04.068. 3

[32]   Ge, S., Zhang, L., and Yu, J. Paper-based microfluidic devices in bioanalysis: how far have we come? *Bioanalysis*. 2015; 7(6):633–636. DOI: 10.4155/bio.15.3. 3

[33]   Pittet, P., Lu, G. N., Galvan, J. M., Ferrigno, R., Blum, L. J., and Leca-Bouvier, B. D. PCB technology-based electrochemiluminescence microfluidic device for low-cost portable analytical systems. *IEEE Sensors Journal.* 2008; 8(5):565–571. DOI: 10.1109/JSEN.2008.918994. 3

[34]   Flores, G., Aracil, C., Perdigones, F., and Quero, J. M. Low consumption single-use microvalve for microfluidic PCB-based platforms. *Journal of Micromechanics and Microengineering.* 2014; 24(6):065013. DOI: 10.1088/0960-1317/24/6/065013. 3

[35]   Tomazelli Coltro, W. K., Cheng, C. M., Carrilho, E, de Jesus, D. P. Recent advances in low-cost microfluidic platforms for diagnostic applications. *Electrophoresis.* 2014; 35(16):2309–2324. DOI: 10.1002/elps.201400006. 3

[36]   Babikian, S., Li, G. P., and Bachman, M. A digital signal processing-assisted microfluidic PCB for on-chip fluorescence detection. *IEEE Transactions on Components, Packaging and Manufacturing Technology.* 2017; 7(6):846–854. DOI: 10.1109/TCPMT.2017.2691673. 3

[37]   Nam, Y., Kim, M., and Kim, T. Fabricating a multi-level barrier-integrated microfluidic device using grey-scale photolithography. *Journal of Micromechanics and Microengineering.* 2013; 23(10):105015. DOI: 10.1088/0960-1317/23/10/105015. 4

[38]   Kim, S. C., Sukovich, D. J., and Abate, A. R. Patterning microfluidic device wettability with spatially-controlled plasma oxidation. *Lab on a Chip.* 2015; 15(15):3163–3169. DOI: 10.1039/C5LC00626K. 4

[39]   Yagüe, J. L., Coclite, A. M., Petruczok, C., and Gleason, K. K. Chemical vapor deposition for solvent-free polymerization at surfaces. *Macromolecular Chemistry and Physics.* 2013; 214(3):302–312. DOI: 10.1002/macp.201200600. 4

[40]   Schulze, M. and Belder, D. Poly(ethylene glycol)-coated microfluidic devices for chip electrophoresis. *Electrophoresis.* 2012; 33(2):370–378. DOI: 10.1002/elps.201100401. 4

[41]   Vlachopoulou, M. E., Kokkoris, G., Cardinaud, C., Gogolides, E., and Tserepi, A. Plasma Etching of poly(dimethylsiloxane): roughness formation, mechanism, control, and application in the fabrication of microfluidic structures. *Plasma Processes and Polymers.* 2013; 10(1):29–40. DOI: 10.1002/ppap.201200008. 4

[42]   Lehmkuhl, B., Noblitt, S. D., Krummel, A. T., and Henry, C. S. Fabrication of IR-transparent microfluidic devices by anisotropic etching of channels in CaF_2. *Lab on a Chip.* 2015; 15(22):4364–4368. DOI: 10.1039/C5LC00759C. 4

[43]   Schomburg, W. K, Vollmer, J., Bustgens, B., Fahrenberg, J., Hein, H., and Menz, W. Microfluidic components in LIGA technique. *Journal of Micromechanics and Microengineering.* 1994; 4(4):186–191. DOI: 10.1088/0960-1317/4/4/003. 4

[44]   Goldenberg, B. G., Goryachkovskaya, T. N., Eliseev, V. S., Kolchanov, N. A., Kondrat'ev, V. I., Kulipanov, G. N., Popik, V. M., Pel'tek, S. E., Petrova, E. V., and Pindyurin, V. F. Fabrication of LIGA masks for microfluidic analytical systems. *Journal of Surface Investigation. X-ray, Synchrotron and Neutron Techniques*. 2008; 2(4):637–640. DOI: 10.1134/S1027451008040265. 4

[45]   Chan, H. N., Chen, Y., Shu, Y., Chen, Y., Tian, Q., and Wu, H. Direct, one-step molding of 3D-printed structures for convenient fabrication of truly 3D PDMS microfluidic chips. *Microfluidics and Nanofluidics*. 2015; 19(1):9–18. DOI: 10.1007/s10404-014-1542-4. 4

[46]   Hou, H. H., Wang, Y. N., Chang, C. L., Yang, R. J., and Fu, L. M. Rapid glucose concentration detection utilizing disposable integrated microfluidic chip. *Microfluidics and Nanofluidics*. 2011; 11(4):479–487. DOI: 10.1007/s10404-011-0813-6. 4

[47]   Joye, C. D., Calame, J. P., Nguyen, K. T., and Garven, M. Microfabrication of fine electron beam tunnels using UV-LIGA and embedded polymer monofilaments for vacuum electron devices. *Journal of Micromechanics and Microengineering*. 2012; 22(1):015010. DOI: 10.1088/0960-1317/22/1/015010. 4

[48]   Unger, M. A. Monolithic Microfabricated valves and pumps by multilayer soft lithography. *Science*. 2000; 288(5463):113–116. DOI: 10.1126/science.288.5463.113. 4

[49]   Love, J. C., Anderson, J. R., and Whitesides, G. M. Fabrication of 3D microfluidic systems by soft lithography. *MRS Bulletin*. 2011; 26(07):523–528. DOI: 10.1557/mrs2001.124. 4

[50]   Huang, L., Bian, S., Cheng, Y., Shi, G., Liu, P., Ye, X., and Wang, W. Microfluidics cell sample preparation for analysis: Advances in efficient cell enrichment and precise single cell capture, *Biomicrofluidics*, 2017, 11(1), 011501, DOI: 10.1063/1.4975666. 5

[51]   Cheng, Y., Wang, Y., Ma, A., Wang, W., and Ye, X. A bubble- and clogging-free microfluidic particle separation platform with multi-filtration, *Lab on a Chip*, 2016, 4517–4526, DOI: 10.1039/C6LC01113F. 5

[52]   Choi, J. W., Oh, K. W., Han, A., Wijayawardhana, C. A., Lannes, C., Bhansali, S., Schlueter, K. T., Heineman, W. R., Halsall, H. B., Nevin, J. H., Helmicki, A. J., Henderson, H. T., and Ahn, C. H. Development and characterization of microfluidic devices and systems for magnetic bead-based biochemical detection. *Biomedical Microdevices*. 2001; 3(3):191–200. DOI: 10.1023/A:1011490627871. 5

[53]   Zhang, S., Ma, Z., Zhang, Y., Wang, Y., Cheng, Y., Wang, W., and Ye, X. On-chip immuno-agglutination assay based on a dynamic magnetic bead clump and a sheath-less flow cytometry, *Biomicrofluidics*, 2019, 13(4), 044102, DOI: 10.1063/1.5093766. 5

[54]   Fisher, K. A., and Miles, R. Modeling the acoustic radiation force in microfluidic chambers. *The Journal of the Acoustical Society of America*. 2008; 123(4):1862–1865. DOI: 10.1121/1.2839140. 5

[55]   Zeng, Q., Guo, F., Yao, L., Zhu, H. W., Zheng, L., Guo, Z. X., Liu, W., Chen, Y., Guo, S. S., and Zhao, X. Z. Milliseconds mixing in microfluidic channel using focused surface acoustic wave. *Sensors and Actuators B: Chemical*. 2011; 160(1):1552–1556. DOI:10.1016/j.snb.2011.08.075. 5

[56]   Kerbage, C. and Eggleton, B. J. Tunable microfluidic optical fiber gratings. *Applied Physics Letters*. 2003; 82(9):1338–1340. DOI: 10.1063/1.1557334. 5

[57]   Tu, L., Li, X., Bian, S., Yu, Y., Li, J., Huang, L., Liu, P., Wu, Q., and Wang, W. Label-free and real-time monitoring of single cell attachment on template-stripped plasmonic nano-holes, *Scientific Reports*, 2017, Article number: 11020. DOI:10.1038/s41598-017-11383-x. 5

[58]   Çetin, B. and Li, D. DEP in microfluidics technology. *Electrophoresis*. 2011; 32(18):2410–2427. DOI: 10.1002/elps.201100167. 5

[59]   Mathew, B., Alazzam, A., Destgeer, G., and Sung, H. J. DEP based cell switching in continuous flow microfluidic devices. *Journal of Electrostatics*. 2016; 84:63–72. DOI: 10.1016/j.elstat.2016.09.003. 5

[60]   Di Carlo, D., Aghdam, N., and Lee, L. P. Single-cell enzyme concentrations, kinetics, and inhibition analysis using high-density hydrodynamic cell isolation arrays. *Analytical Chemistry*. 2006; 78(14):4925–4930. DOI: 10.1021/ac060541s. 5

[61]   Tan, W. H. and Takeuchi, S. A trap-and-release integrated microfluidic system for dynamic microarray applications. *Proceedings of the National Academy of Sciences*. 2007; 104(4):1146–1151. DOI: 10.1073/pnas.0606625104. 5

[62]   Tallapragada, P., Hasabnis, N., Katuri, K., Sudarsanam, S., Joshi, K,. and Ramasubramanian, M. Scale invariant hydrodynamic focusing and sorting of inertial particles by size in spiral micro channels. *Journal of Micromechanics and Microengineering*. 2015; 25(8):084013. DOI: 10.1088/0960-1317/25/8/084013. 5

[63]   Xiang, N. and Ni, Z. High-throughput blood cell focusing and plasma isolation using spiral inertial microfluidic devices. *Biomedical Microdevices*. 2015; 17(6):110. DOI: 10.1007/s10544-015-0018-y. 5

[64]   Zhang, K., Chou, C. K., Xia, X., Hung, M. C., and Qin, L. Block-cell-printing for live single-cell printing. *Proceedings of the National Academy of Sciences*. 2014; 111(8):2948–2953. DOI: 10.1073/pnas.1313661111. 5, 6

[65]   Jin, D., Deng, B., Li, J. X., Cai, W., Tu, L., Chen, J., Wu, Q., and Wang, W. H. A microfluidic device enabling high-efficiency single cell trapping. *Biomicrofluidics*. 2015; 9(1):014101. DOI: 10.1063/1.4905428. 5, 6

[66]   Mi, L., Huang, L., Li, J., Xu, G., Wu, Q., and Wang, W. A fluidic circuit based, high-efficiency and large-scale single cell trap. *Lab on a Chip*. 2016; 16(23):4507–4511. DOI: 10.1039/C6LC01120A. 5, 6

[67]   Bhagat, A. A. S., Kuntaegowdanahalli, S. S., and Papautsky, I. Continuous particle separation in spiral microchannels using dean flows and differential migration. *Lab on a Chip*. 2008; 8(11):1906–1914. DOI: 10.1039/b807107a. 6

[68]   Warkiani, M. E., Guan, G., Luan, K. B., Lee, W. C., Bhagat, A. A. S., Kant Chaudhuri, P., Tan, D. S-W., Lim, W. T., Lee, S. C., Chen, P. C. Y., Lim, C. T., and Han, J. Slanted spiral microfluidics for the ultra-fast, label-free isolation of circulating tumor cells. *Lab Chip*. 2014; 14(1):128–137. DOI: 10.1039/C3LC50617G. 6

[69]   Liu, Z., Huang, F., Du, J., Shu, W., Feng, H., Xu, X., and Chen, Y. Rapid isolation of cancer cells using microfluidic deterministic lateral displacement structure. *Biomicrofluidics*. 2013; 7(1):011801. DOI: 10.1063/1.4774308. 6

[70]   Okano, H., Konishi, T., Suzuki, T., Suzuki, T., Ariyasu, S., Aoki, S., Abe, R., and Hayase, M. Enrichment of circulating tumor cells in tumor-bearing mouse blood by a deterministic lateral displacement microfluidic device. *Biomedical Microdevices*. 2015; 17(3):59. DOI: 10.1007/s10544-015-9964-7. 6

[71]   Al-Halhouli, A., Albagdady, A., Al-Faqheri, W., Kottmeier, J., Meinen, S., Frey, L. J., Krull, R., and Dietzel, A.. Enhanced inertial focusing of microparticles and cells by integrating trapezoidal microchambers in spiral microfluidic channels. *RSC Advances*. 2019; 9:19197–19204. DOI: 10.1039/C9RA03587G. 6, 7

[72]   Liu, Z., Zhang, W., Huang, F., Feng, H., Shu, W., Xu, X., and Chen, Y. High throughput capture of circulating tumor cells using an integrated microfluidic system. *Biosensors and Bioelectronics*. 2013; 47:113–119. DOI: 10.1016/j.bios.2013.03.017. 6, 7

[73]   Ma, B., Yao, B., Peng, F., Yan, S., Lei, M., and Rupp, R. Optical sorting of particles by dual-channel line optical tweezers. *Journal of Optics*. 2012; 14(10):105702. DOI: 10.1088/2040-8978/14/10/105702. 7

[74]   Brzobohatý, O., Karásek, V., Šiler, M., Chvátal, L., Čižmár, T., and Zemánek, P.. Experimental demonstration of optical transport, sorting and self-arrangement using a "tractor beam." *Nature Photonics*. 2013; 7(2):123–127. DOI: 10.1038/nphoton.2012.332. 7

[75]   Chen, H. and Sun, D. Moving groups of microparticles into array with a robot–twee-zers manipulation system. *IEEE Transactions on Robotics*. 2012; 28(5):1069–1080. DOI: 10.1109/TRO.2012.2196309. 7

[76]   Yan, X. and Sun, D. Multilevel-based topology design and cell patterning with roboti-cally controlled optical tweezers. *IEEE Transactions on Control Systems Technology*. 2015; 23(1):176–185. DOI: 10.1109/TCST.2014.2317798. 7

[77]   Liberale, C., Cojoc, G., Bragheri, F., Minzioni, P., Perozziello, G., La Rocca, R., Ferrara, L., Rajamanickam, V., Di Fabrizio, E., and Cristiani, I. Integrated microfluidic device for single-cell trapping and spectroscopy. *Scientific Reports*. 2013; 3(1):1258. DOI: 10.1038/srep01258. 7, 8

[78]   Werner, M., Merenda, F., Piguet, J., Salathé, R-P., and Vogel, H. Microfluidic array cy-tometer based on refractive optical tweezers for parallel trapping, imaging and sorting of individual cells. *Lab on a Chip*. 2011; 11(14):2432–2439. DOI: 10.1039/c1lc20181f. 7, 8

[79]   Landenberger, B., Höfemann, H., Wadle, S., and Rohrbach, A. Microfluidic sorting of arbitrary cells with dynamic optical tweezers. *Lab on a Chip*. 2012; 12(17):3177–3183. DOI: 10.1039/c2lc21099a. 7, 8

[80]   Wang, X., Chen, S., Kong, M., Wang, Z., Costa, K. D., Li, R. A., and Sun, D. Enhanced cell sorting and manipulation with combined optical tweezer and microfluidic chip tech-nologies. *Lab on a Chip*. 2011; 11(21):3656–3662. DOI: 10.1039/c1lc20653b. 7, 8

[81]   He, R., Zhao, L., Liu, Y., Zhang, N., Cheng, B., He, Z., Cai, B., Li, S., Liu, W., Guo, S., Chen, Y., Xiong, B., and Zhao, X. Z. Biocompatible TiO_2 nanoparticle-based cell immunoassay for circulating tumor cells capture and identification from cancer patients. *Biomedical Microdevices*. 2013; 15(4):617–626. DOI: 10.1007/s10544-013-9781-9. 8

[82]   Hyun, K-A., Lee, T. Y., and Jung, H. I. Negative enrichment of circulating tumor cells using a geometrically activated surface interaction chip. *Analytical Chemistry*. 2013; 85(9):4439–4445. DOI: 10.1021/ac3037766. 8

[83]   Chen, P., Huang, Y. Y., Bhave, G., Hoshino, K., and Zhang, X. Inkjet-print micromagnet array on glass slides for immunomagnetic enrichment of circulating tumor cells. *Annals of Biomedical Engineering*. 2015; 44(5):1710–1720. DOI: 10.1007/s10439-015-1427-z. 8

[84]   Lu, Y., Liang, H., Yu, T., Xie, J., Chen, S., Dong, H., Sinko, P. J., Lian, S., Xu, J., Wang, J., Yu, S., Shao, J., Yuan, B., Wang, L., and Jia, L. Isolation and characterization of living circulating tumor cells in patients by immunomagnetic negative enrichment coupled with flow cytometry. *Cancer*. 2015; 121(17):3036–3045. DOI: 10.1002/cncr.29444. 8

[85] Huang, Y. Y., Chen, P., Wu, C. H., Hoshino, K., Sokolov, K., Lane, N., Liu, H., Huebschman, M., Frenkel, E., and Zhang, J. X. J. Screening and molecular analysis of single circulating tumor cells using micromagnet array. *Scientific Reports*. 2015; 5(1):16047. DOI: 10.1038/srep16047. 8, 9

[86] Ding, X., Peng, Z., Lin, S. C. S., Geri, M., Li, S., Li, P., Chen, Y., Dao, M., Suresh, S., and Huang, T. J. Cell separation using tilted-angle standing surface acoustic waves. *Proceedings of the National Academy of Sciences*. 2014; 111(36):12992–12997. DOI: 10.1073/pnas.1413325111. 9

[87] Devendran, C., Gunasekara, N. R., Collins, D. J., and Neild, A. Batch process particle separation using surface acoustic waves (SAW): integration of travelling and standing SAW. *RSC Advances*. 2016; 6(7):5856–5864. DOI: 10.1039/C5RA26965B. 9

[88] Guo, F., Mao, Z., Chen, Y., Xie, Z., Lata, J. P., Li, P., Ren, L., Liu, J., Yang, J., Dao, M., Suresh, S., and Huang, T. J. 3D manipulation of single cells using surface acoustic waves. *Proceedings of the National Academy of Sciences*. 2016; 113(6):1522–1527. DOI: 10.1073/pnas.1524813113. 9

[89] Li, P., Mao, Z., Peng, Z., Zhou, L., Chen, Y., Huang, P-H., Truica, C. I., Drabick, J. J., El-Deiry, W. S., Dao, M., Suresh, S., and Huang, T. J. Acoustic separation of circulating tumor cells. *Proceedings of the National Academy of Sciences*. 2015; 112(16):4970–4975. DOI: 10.1073/pnas.1504484112. 9, 10

[90] Collins, D. J., Neild, A., and Ai, Y. Highly focused high-frequency travelling surface acoustic waves (SAW) for rapid single-particle sorting. *Lab on a Chip*. 2016; 16(3):471–479. DOI: 10.1039/C5LC01335F. 9, 10

[91] Urdaneta, M. and Smela, E. Parasitic trap cancellation using multiple frequency DEP, demonstrated by loading cells into cages. *Lab on a Chip*. 2008; 8(4):550–556. DOI: 10.1039/b717862j. 10

[92] Chen, N. C., Chen, C. H., Chen, M. K., Jang, L. S., and Wang, M. H. Single-cell trapping and impedance measurement utilizing DEP in a parallel-plate microfluidic device. *Sensors and Actuators B: Chemical*. 2014; 190:570–577. DOI: 10.1016/j.snb.2013.08.104. 10

[93] He, W., Huang, L., Feng, Y., Liang, F., Ding, W., and Wang, W. Highly integrated microfluidic device for cell pairing, fusion and culture, *Biomicrofluidics*, 2019, 13(5), 054109. DOI: 10.1063/1.5124705. 10

[94] Kimura, Y., Gel, M., Techaumnat, B., Oana, H., Kotera, H., and Washizu, M. DEP-assisted massively parallel cell pairing and fusion based on field constriction created by

a micro-orifice array sheet. *Electrophoresis.* 2011; 32(18):2496–2501. DOI: 10.1002/elps.201100129. 10

[95] Chen, D. F., Du, H., and Li, W. H. Bioparticle separation and manipulation using DEP. *Sensors and Actuators A: Physical.* 2007; 133(2):329–334. DOI: 10.1016/j.sna.2006.06.029. 10

[96] Jones, P. V., Salmon, G. L., and Ros, A. Continuous separation of DNA molecules by size using insulator-based DEP. *Analytical Chemistry.* 2017; 89(3):1531–1539. DOI: 10.1021/acs.analchem.6b03369. 10

[97] Song, H., Rosano, J. M., Wang, Y., Garson, C. J., Prabhakarpandian, B., Pant, K., Klarmann, G. J., Perantoni, A., Alvarez, L. M., and Lai, E. Continuous-flow sorting of stem cells and differentiation products based on DEP. *Lab on a Chip.* 2015; 15(5):1320–1328. DOI: 10.1039/C4LC01253D. 10, 11

[98] Gallo-Villanueva, R. C., Rodríguez-López, C, E., Díaz-de-la-Garzma, R. I., Reyes-Betanzo, C., and Lapizco-Encinas, B. H. DNA manipulation by means of insulator-based DEP employing direct current electric fields. *Electrophoresis.* 2009; 30(24):4195–4205. DOI: 10.1002/elps.200900355. 10

[99] Nakidde, D., Zellner, P., Alemi, M. M., Shake, T., Hosseini, Y., Riquelme, M. V, Pruden, A., and Agah, M. Three dimensional passivated-electrode insulator-based DEP. *Biomicrofluidics.* 2015; 9(1):014125. DOI: 10.1063/1.4913497. 10

[100] LaLonde, A., Romero-Creel, M. F., Saucedo-Espinosa, M. A., and Lapizco-Encinas, B. H. Isolation and enrichment of low abundant particles with insulator-based DEP. *Biomicrofluidics.* 2015; 9(6):064113. DOI: 10.1063/1.4936371. 10, 11

[101] Jones, T. B. Basic theory of DEP and electrorotation. *IEEE Engineering in Medicine and Biology Magazine.* 2003; 22(6):33–42. DOI: 10.1109/MEMB.2003.1304999. 12

[102] Pohl, H. A., Pollock, K., and Crane, J. S. DEP force: A comparison of theory and experiment. *Journal of Biological Physics.* 1978; 6(3-4):133–160. DOI: 10.1007/BF02328936. 13

[103] Jones, T. B. and Washizu, M. Multipolar DEP and electrorotation theory. *Journal of Electrostatics.* 1996; 37(1-2):121–134. DOI: 10.1016/0304-3886(96)00006-X. 14

[104] Henning, A., Bier, F. F., and Hölzel, R. DEP of DNA: Quantification by impedance measurements. *Biomicrofluidics.* 2010;4(2):022803. DOI: 10.1063/1.3430550. 15

[105] Nakano, M., Ding, Z., and Suehiro, J. DEP and DEP impedance detection of adenovirus and rotavirus. *Japanese Journal of Applied Physics.* 2016; 55(1):017001. DOI: 10.7567/JJAP.55.017001. 15

[106] Suehiro, J., Noutomi, D., Shutou, M., and Hara, M. Selective detection of specific bacteria using DEP impedance measurement method combined with an antigen–antibody reaction. *Journal of Electrostatics*. 2003; 58(3-4):229–246. DOI: 10.1016/S0304-3886(03)00062-7. 15

[107] Suehiro, J., Ohtsubo, A., Hatano, T., and Hara, M. Selective detection of bacteria by a DEP impedance measurement method using an antibody-immobilized electrode chip. *Sensors and Actuators B: Chemical*. 2006; 119(1):319–326. DOI: 10.1016/j.snb.2005.12.027. 15

[108] Huang, L., Tu, L., Zeng, X., Mi, L., Li, X., and Wang, W. Study of a microfluidic chip integrating single cell trap and 3D stable rotation manipulation, *Micromachines*, 2016, 7(8), 141. DOI: 10.3390/mi7080141. 15

[109] Zhang, P., Ren, L., Zhang, X., Shan, Y., Wang, Y., Ji, Y., Yin, H., Huang, W. E., Xu, J., and Ma, B. Raman-activated cell sorting based on DEP single-cell trap and release. *Analytical Chemistry*. 2015; 87(4):2282–2289. DOI: 10.1021/ac503974e. 15

[110] Tu, L., Huang, L., and Wang, W. A novel micromachined Fabry-Perot interferometer integrating nano-holes and dielectrophoresis for enhanced biochemical sensing, Biosensors and Bioelectronics, 2019, 127, 19-24. DOI: 10.1016/j.bios.2018.12.013. 15

[111] Gonzalez-Moragas, L., Roig, A., and Laromaine, A. C. elegans as a tool for in vivo nanoparticle assessment. *Advances in Colloid and Interface Science*. 2015; 219:10–26. DOI: 10.1016/j.cis.2015.02.001.

[112] Edison, A., Hall, R., Junot, C., Karp, P., Kurland, I., Mistrik, R., Reed, L., Saito, K., Salek, R., Steinbeck, C., Sumner, L., and Viant, M. The time is right to focus on model organism metabolomes. *Metabolites*. 2016; 6(1):8. DOI: 10.3390/metabo6010008.

[113] Bradham, C. A., Foltz, K. R., Beane, W. S., Arnone, M. I., Rizzo, F., Coffman, J. A., Mushegian, A., Goel, M., Morales, J., Geneviere, A. M., Lapraz, F., Robertson, A. J., Kelkar, H., Loza-Coll, M., Townley, I. K., Raisch, M., Roux, M. M., Lepage, T., Gache, C., McClay, D. R., and Manning, G. The sea urchin kinome: A first look. *Developmental Biology*. 2006; 300(1):180–193. DOI: 10.1016/j.ydbio.2006.08.074.

[114] Clavería, C. and Torres, M. Mitochondrial apoptotic pathways induced by drosophila programmed cell death regulators. *Biochemical and Biophysical Research Communications*. 2003; 304(3):531–537. DOI: 10.1016/S0006-291X(03)00626-0.

[115] Litke, R., Boulanger, É., and Fradin, C. Caenorhabditis elegans, un modèle d'étude du vieillissement. *Médecine/Sciences*. 2018; 34(6-7):571-579. DOI: 10.1051/medsci/20183406017.

[116]   Kaeberlein, M., Burtner, C. R., and Kennedy, B. K. Recent developments in yeast aging. *PLoS Genetics*. 2007; 3(5):e84. DOI: 10.1371/journal.pgen.0030084.

[117]   Hwang, H. and Lu, H. Microfluidic tools for developmental studies of small model organisms—nematodes, fruit flies, and zebrafish. *Biotechnology Journal*. 2013; 8(2):192–205. DOI: 10.1002/biot.201200129.

[118]   Park, K-W. and Li, L. Cytoplasmic expression of mouse prion protein causes severe toxicity in Caenorhabditis elegans. *Biochemical and Biophysical Research Communications*. 2008; 372(4):697–702. DOI: 10.1016/j.bbrc.2008.05.132.

[119]   Lee, H., Kim, S. A., Coakley, S., Mugno, P., Hammarlund, M., Hilliard, M. A., and Lu, H. A multi-channel device for high-density target-selective stimulation and long-term monitoring of cells and subcellular features in C. elegans. *Lab on a Chip*. 2014; 14(23):4513–4522. DOI: 10.1039/C4LC00789A.

[120]   Cho, Y., Porto, D. A., Hwang, H., Grundy, L. J., Schafer, W. R., and Lu, H. Automated and controlled mechanical stimulation and functional imaging in vivo in C. elegans. *Lab on a Chip*. 2017;17(15):2609–2618. DOI: 10.1039/C7LC00465F.

[121]   Gupta, B. and Rezai, P. Microfluidic approaches for manipulating, imaging, and screening C. elegans. *Micromachines*. 2016; 7(7):123. DOI: 10.3390/mi7070123.

[122]   Kim, W., Hendricks, G. L., Lee, K., and Mylonakis, E. An update on the use of C. elegans for preclinical drug discovery: screening and identifying anti-infective drugs. *Expert Opinion on Drug Discovery*. 2017; 12(6):625–633. DOI: 10.1080/17460441.2017.1319358.

[123]   Khoshmanesh, K., Kiss, N., Nahavandi, S., Evans, C. W., Cooper, J. M., Williams, D. E., and Wlodkowic, D. Trapping and imaging of micron-sized embryos using DEP. *Electrophoresis*. 2011; 32(22):3129–3132. DOI: 10.1002/elps.201100160.

[124]   Chuang, H-S., Raizen, D. M., Lamb, A., Dabbish, N., and Bau, H. H. Dielectrophoresis of Caenorhabditis elegans. *Lab on a Chip*. 2011; 11(4):599–604. DOI: 10.1039/c0lc00532k.

[125]   Gavrieli, Y. Identification of programmed cell death in situ via specific labeling of nuclear DNA fragmentation. *The Journal of Cell Biology*. 1992; 119(3):493–501. DOI: 10.1083/jcb.119.3.493. 15

[126]   Anchang, B., Davis, K. L., Fienberg, H. G., Williamson, B. D., Bendall, S. C., Karacosta, L. G., Tibshirani, R., Nolan, G. P., and Plevritis, S. K. DRUG-NEM: Optimizing drug combinations using single-cell perturbation response to account for intratumoral heterogeneity. *Proceedings of the National Academy of Sciences*. 2018; 115(18), E4294-E4303. DOI: 10.1073/pnas.1711365115. 15

[127]   Steinert, G., Schölch, S., Niemietz, T., Iwata, N., García, S. A., Behrens, B., Voigt, A., Kloor, M., Benner, A., Bork, U., Rahbari, N. N., Büchler, M. W., Stoecklein, N. H., Weitz, J., and Koch, M. Immune escape and survival mechanisms in circulating tumor cells of colorectal cancer. *Cancer Research*. 2014; 74(6):1694–1704. DOI: 10.1158/0008-5472. CAN-13-1885. 15

[128]   Paoletti, C., Muñiz, M. C., Thomas, D. G., Griffith, K. A., Kidwell, K. M., Tokudome, N., Brown, M. E., Aung, K., Miller, M. C., Blossom, D. L., Schott, A. F., Henry, N. L., Rae, J. M., Connelly, M. C., Chianese, D. A., and Hayes, D. F. Development of circulating tumor cell-endocrine therapy index in patients with hormone receptor–positive breast cancer. *Clinical Cancer Research*. 2015; 21(11):2487–2498. DOI: 10.1158/1078-0432.CCR-14-1913. 16

[129]   Pethig, R. Review—Where is DEP (DEP) going? *Journal of The Electrochemical Society*. 2016; 164(5):B3049–B3055. DOI: 10.1149/2.0071705jes. 16

[130]   Kung, Y. C., Huang, K. W., Chong, W., and Chiou, P. Y. Tunnel DEP for tunable, single-stream cell focusing in physiological buffers in high-speed microfluidic flows. *Small*. 2016; 12(32):4343–4348. DOI: 10.1002/smll.201600996. 16

[131]   Jang, L. S., Huang, P. H., and Lan, K. C. Single-cell trapping utilizing negative DEP quadrupole and microwell electrodes. *Biosensors and Bioelectronics*. 2009; 24(12):3637–3644. DOI: 10.1016/j.bios.2009.05.027. 16

[132]   Guo, X. and Zhu, R. Controllably moving individual living cell in an array by modulating signal phase shift based on DEP. *Biosensors and Bioelectronics*. 2015; 68:529–535. DOI: 10.1016/j.bios.2015.01.052. 16

[133]   Kemna, E. W. M., Wolbers, F., Vermes, I., and van den Berg, A. On chip electrofusion of single human B cells and mouse myeloma cells for efficient hybridoma generation. *Electrophoresis*. 2011; 32(22):3138–3146. DOI: 10.1002/elps.201100227. 16

[134]   Huang, L., He, W., and Wang, W. A cell electro-rotation micro-device using polarized cells as electrodes, *Electrophoresis, Special Issue: Fundamentals* 2019, 2019, 40(5), 784-791. DOI: 10.1002/elps.201800360. 16

[135]   Wu, Y., Ren, Y., Tao, Y., Hou, L., and Jiang, H. High-throughput separation, trapping, and manipulation of single cells and particles by combined dielectrophoresis at a bipolar electrode array. *Analytical Chemistry*. 2018; 90(19):11461–11469. DOI: 10.1021/acs. analchem.8b02628. 16, 17

[136]   Şen, M., Ino, K., Ramón-Azcón, J., Shiku. H,, and Matsue, T. Cell pairing using a DEP-based device with interdigitated array electrodes. *Lab on a Chip*. 2013; 13(18):3650–3652. DOI: 10.1039/c3lc50561h. 16, 17

[137]   Gel, M., Kimura, Y., Kurosawa, O., Oana, H., Kotera, H., and Washizu, M. DEP cell trapping and parallel one-to-one fusion based on field constriction created by a micro-orifice array. *Biomicrofluidics*. 2010; 4(2):022808. DOI: 10.1063/1.3422544. 16, 17

[138]   Bahrieh, G., Erdem, M., Özgür, E., Gündüz, U., and Külah, H. Assessment of effects of multi drug resistance on dielectric properties of K562 leukemic cells using electrorotation. *RSC Adv*. 2014;4(85):44879–44887. DOI: 10.1039/C4RA04873C. 16, 17

[139]   Zeinali, S., Çetin, B., Oliaei, S. N. B., and Karpat, Y. Fabrication of continuous flow microfluidics device with 3D electrode structures for high throughput DEP applications using mechanical machining. *Electrophoresis*. 2015;36(13):1432–1442. DOI: 10.1002/elps.201400486. 18

[140]   Voldman, J., Gray, M. L., Toner, M., and Schmidt, M. A. A microfabrication-based dynamic array cytometer. *Analytical Chemistry*. 2002; 74(16):3984–3990. DOI: 10.1021/ac0256235. 19

[141]   So, J. H. and Dickey, M. D. Inherently aligned microfluidic electrodes composed of liquid metal. *Lab on a Chip*. 2011; 11(5), 905–911. DOI: 10.1039/c0lc00501k. 19

[142]   Elvington, E. S., Salmanzadeh, A., Stremler, M. A., and Davalos, R. V. Label-free isolation and enrichment of cells through contactless DEP. *Journal of Visualized Experiments*. 2013(79): e50634. DOI: 10.3791/50634. 19

[143]   Čemažar, J., Douglas, T. A., Schmelz, E. M., and Davalos, R. V. Enhanced contactless DEP enrichment and isolation platform via cell-scale microstructures. *Biomicrofluidics*. 2016; 10(1):014109. DOI: 10.1063/1.4939947. 19

[144]   Shafiee, H., Caldwell, J. L., and Davalos, R. V. A microfluidic system for biological particle enrichment using contactless dielectrophoresis. *Journal of Laboratory Automation*. 2010; 15(3):224–232. DOI: 10.1016/j.jala.2010.02.003. 19

[145]   Huang, L., Zhao, P., and Wang, W. 3D cell electrorotation and imaging for measuring multiple cellular biophysical properties, *Lab on a Chip*, 2018, 18, 2359-2368. DOI: 10.1039/C8LC00407B. 20

[146]   Lewpiriyawong, N. and Yang, C. AC-DEP characterization and separation of submicron and micron particles using sidewall AgPDMS electrodes. *Biomicrofluidics*. 2012; 6(1):012807. DOI: 10.1063/1.3682049. 20

[147] Marchalot, J., Chateaux, J-F., Faivre, M., Mertani, H. C., Ferrigno, R., and Deman, A-L. DEP capture of low abundance cell population using thick electrodes. *Biomicrofluidics*. 2015; 9(5):054104. DOI: 10.1063/1.4928703. 20

[148] Deman, A. L., Brun, M., Quatresous, M., Chateaux, J. F., Frenea-Robin, M., Haddour, N., Semet, V., and Ferrigno, R. Characterization of C-PDMS electrodes for electroki- netic applications in microfluidic systems. *Journal of Micromechanics and Microengineering*. 2011; 21(9):095013. DOI: 10.1088/0960-1317/21/9/095013. 23

[149] Huang, L., Zhao, P., Wu, J., Chuang, H-S., and Wang, W. On-demand dielectrophoretic immobilization and high-resolution imaging of C. elegans in microfluids, *Sensors and Actuators B: Chemical*, 2018, 259, 703-708. DOI: 10.1016/j.snb.2017.12.106. 23

[150] Wang, L., Flanagan, L. A., Jeon, N. L., Monuki, E., and Lee, A. P. Dielectrophoresis switching with vertical sidewall electrodes for microfluidic flow cytometry. *Lab on a Chip*. 2007; 7(9):1114–1120. DOI: 10.1039/b705386j. 23

[151] Li, S., Li, M., Hui, Y. S., Cao, W., Li, W., and Wen, W. A novel method to construct 3D electrodes at the sidewall of microfluidic channel. *Microfluidics and Nanofluidics*. 2012; 14(3-4):499–508. DOI: 10.1007/s10404-012-1068-6. 23

[152] Kang, Y., Cetin, B., Wu, Z., and Li, D. Continuous particle separation with localized AC-dielectrophoresis using embedded electrodes and an insulating hurdle. *Electrochimica Acta*. 2009; 54(6):1715–1720. DOI: 10.1016/j.electacta.2008.09.062. 23

[153] Miao, Q., Reeves, A. P., Patten, F. W., and Seibel, E. J. Multimodal 3D imaging of cells and tissue, bridging the gap between clinical and research microscopy. *Annals of Biomed- ical Engineering*. 2011; 40(2):263–276. DOI: 10.1007/s10439-011-0411-5. 24

[154] Habaza, M., Kirschbaum, M., Guernth-Marschner, C., Dardikman, G., Barnea, I., Ko- renstein, R., Duschl, C., and Shaked, N. T. Rapid 3D refractive-index imaging of live cells in suspension without labeling using DEP cell rotation. *Advanced Science*. 2017; 4(2):1600205. DOI: 10.1002/advs.201600205. 24

[155] Chen, D., Sun, M., and Zhao, X. Oocytes polar body detection for automatic enucleation. *Micromachines*. 2016; 7(2):27. DOI: 10.3390/mi7020027. 25

[156] Bin, C., Kelbauskas, L., Chan, S., Shetty, R. M., Smith, D., and Meldrum, D. R. Rotation of single live mammalian cells using dynamic holographic optical tweezers. *Optics and Lasers in Engineering*. 2017; 92:70–75. DOI: 10.1016/j.optlaseng.2016.12.019. 25

[157] Merola, F., Miccio, L., Memmolo, P. , Di Caprio, G., Galli, A., Puglisi, R., Balduzzi, D., Coppola, G., Netti, P., and Ferraro, P. Digital holography as a method for 3D imaging

and estimating the biovolume of motile cells. *Lab on a Chip*. 2013; 13(23):4512–4516. DOI: 10.1039/c3lc50515d. 25

[158]   Elbez, R., McNaughton, B. H., Patel, L., Pienta, K. J., and Kopelman, R. Nanoparticle induced cell magneto-rotation: monitoring morphology, stress and drug sensitivity of a suspended single cancer cell. *PLoS ONE*. 2011; 6(12):e28475. DOI: 10.1371/journal. pone.0028475. 25, 26

[159]   Berret, J. F. Local viscoelasticity of living cells measured by rotational magnetic spectroscopy. *Nature Communication*s. 2016; 7:10134. DOI: 10.1038/ncomms10134. 25

[160]   Ahmed, D., Ozcelik, A., Bojanala, N., Nama, N., Upadhyay, A., Chen, Y., Hanna-Rose, W., and Huang, T. J. Rotational manipulation of single cells and organisms using acoustic waves. *Nature Communication*. 2016; 7:11085. DOI: 10.1038/ncomms11085. 25, 26

[161]   Ozcelik, A., Nama, N., Huang, P. H., Kaynak, M., McReynolds, M. R., Hanna-Rose, W., and Huang, T. J. Acoustofluidic rotational manipulation of cells and organisms using oscillating solid structures. Sm*all*. 2016; 12(37):5120–5125. DOI: 10.1002/smll.201670191. 25

[162]   Yalikun, Y., Kanda, Y., and Morishima, K.. Hydrodynamic vertical rotation method for a single cell in an open space. *Microfluidics and Nanofluidics*. 2016;20(5): 74. DOI: 10.1007/ s10404-016-1737-y. 26

[163]   Shelby, J. P. and Chiu, D. T. Controlled rotation of biological micro- and nano-particles in microvortices. *Lab on a Chip*. 2004; 4:168–170. DOI: 10.1039/b402479f. 26

[164]   Shetty, R. M., Myers, J. R., Sreenivasulu, M., Teller, W., Vela, J., Houkal. J., Chao, S. H., Johnson, R. H., Kelbauskas, L., Wang, H., and Meldrum, D. R. Characterization and comparison of three microfabrication methods to generate out-of-plane microvortices for single cell rotation and 3D imaging. *Journal of Micromechanics and Microengineering*. 2017; 27(1):015004. DOI: 10.1088/0960-1317/27/1/015004. 26

[165]   Han, S. I., Joo, Y. D., and Han, K. H. An electrorotation technique for measuring the dielectric properties of cells with simultaneous use of negative quadrupolar DEP and electrorotation. *The Analyst*. 2013; 138(5):1529–1537. DOI: 10.1039/c3an36261b. 27, 50, 52

[166]   Liang, Y. L., Huang, Y. P., Lu, Y. S., Hou, M. T., and Yeh, J. A. Cell rotation using optoelectronic tweezers. *Biomicrofluidics*. 2010; 4(4):043003. DOI: 10.1063/1.3496357. 27

[167]   Benhal, P., Chase, J. G., Gaynor, P., Oback, B., and Wang, W. AC electric field induced dipole-based on-chip 3D cell rotation. *Lab on a Chip*. 2014; 14(15):2717–2727. DOI: 10.1039/C4LC00312H. 28

[168] Lei, U., Sun, P. H., and Pethig, R. Refinement of the theory for extracting cell dielectric properties from DEP and electrorotation experiments. *Biomicrofluidics*. 2011; 5(4):044109. DOI: 10.1063/1.3659282. 48

[169] Trainito, C. I., Bayart, E., Subra, F., Français, O., and Le Pioufle, B. The electrorotation as a tool to monitor the dielectric properties of spheroid during the permeabilization. *The Journal of Membrane Biology*. 2016; 249(5):593–600. DOI: 10.1007/s00232-016-9880-7. 48

[170] Huang, C., Chen, A., Guo, M., and Yu, J. Membrane dielectric responses of bufalin-induced apoptosis in HL-60 cells detected by an electrorotation chip. *Biotechnology Letters*. 2007; 29(9):1307–1313. DOI: 10.1007/s10529-007-9414-6. 50

[171] Hamoud Al-Tamimi, M. S., Sulong, G., and Shuaib, I. L. Alpha shape theory for 3D visualization and volumetric measurement of brain tumor progression using magnetic resonance images. *Magnetic Resonance Imaging*. 2015; 33(6):787–803. DOI: 10.1016/j. mri.2015.03.008. 53

[172] Haselgrübler, T., Haider, M., Ji, B., Juhasz, K., Sonnleitner, A., Balogi, Z., and Hesse, J. High-throughput, multiparameter analysis of single cells. *Analytical and Bioanalytical Chemistry*. 2013; 406(14):3279–3296. DOI: 10.1007/s00216-013-7485-x. 57

[173] Hyun, K. A., Kim, J., Gwak, H., and Jung, H. I. Isolation and enrichment of circulating biomarkers for cancer screening, detection, and diagnostics. *Analyst*. 2016; 141(2):382–392. DOI: 10.1039/C5AN01762A. 57

[174] Tang, L. and Casas, J. Quantification of cardiac biomarkers using label-free and multiplexed gold nanorod bioprobes for myocardial infarction diagnosis. *Biosensors and Bioelectronics*. 2014; 61:70–75. DOI: 10.1016/j.bios.2014.04.043. 57

[175] Wang, Q., Liu, F., Yan,g X., Wang, K., Wang, H., and Deng, X. Sensitive point-of-care monitoring of cardiac biomarker myoglobin using aptamer and ubiquitous personal glucose meter. *Biosensors and Bioelectronics*. 2015; 64:161–164. DOI: 10.1016/j. bios.2014.08.079. 57

[176] Nahavandi, S., Baratchi, S., Soffe, R., Tang, S. Y., Nahavandi, S., Mitchell, A., and Khoshmanesh, K. Microfluidic platforms for biomarker analysis. *Lab on a Chip*. 2014; 14(9):1496–1514. DOI: 10.1039/C3LC51124C. 57

[177] Sinha, B., Köster, D., Ruez, R., Gonnord, P., Bastiani, M., Abankwa, D., Stan, R. V., Butler-Browne, G., Vedie, B., Johannes, L., Morone, N., Parton, R. G., Raposo, G., Sens, P., Lamaze, C., and Nassoy, P. Cells Respond to mechanical stress by rapid disassembly of caveolae. *Cell*. 2011; 144(3):402–413. DOI: 10.1016/j.cell.2010.12.031. 57

[178]   Wang, N., Butler, J., and Ingber, D. Mechanotransduction across the cell surface and through the cytoskeleton. *Science*. 1993; 260(5111):1124–1127. DOI: 10.1126/science.7684161. 58

[179]   Chen, L., Maybeck, V., Offenhäusser, A., and Krause, H. J. Implementation and application of a novel 2D magnetic twisting cytometry based on multi-pole electromagnet. *Review of Scientific Instruments*. 2016; 87(6):064301. DOI: 10.1063/1.4954185. 58

[180]   Costa, K. D. Single-cell elastography: probing for disease with the atomic force microscope. *Disease Markers*. 2004; 19(2-3):139–154. DOI: 10.1155/2004/482680. 58, 59

[181]   Wang, K., Cheng, J., Han Cheng, S., and Sun, D. Probing cell biophysical behavior based on actin cytoskeleton modeling and stretching manipulation with optical tweezers. *Applied Physics Letters*. 2013; 103(8):083706. DOI: 10.1063/1.4819392. 58, 59

[182]   Hogan, B., Babataheri, A., Hwang, Y., Barakat Abdul, I., and Husson, J. Characterizing cell adhesion by using micropipette aspiration. *Biophysical Journal*. 2015; 109(2):209–219. DOI: 10.1016/j.bpj.2015.06.015. 58

[183]   Shojaei-Baghini, E., Zheng, Y., and Sun, Y. Automated micropipette aspiration of single cells. *Annals of Biomedical Engineering*. 2013; 41(6):1208–1216. DOI: 10.1007/s10439-013-0791-9. 58

[184]   Worthen, G., Schwab, B., Elson, E., and Downey, G. Mechanics of stimulated neutrophils: cell stiffening induces retention in capillaries. *Science*. 1989; 245(4914):183–186. DOI: 10.1126/science.2749255. 59

[185]   Baskurt, O. K., Gelmont, D., and Meiselman, H. J. Red blood cell deformability in sepsis. *American Journal of Respiratory and Critical Care Medicine*. 1998; 157(2):421–427. DOI: 10.1164/ajrccm.157.2.9611103. 59

[186]   Bow, H., Pivkin, I. V., Diez-Silva, M., Goldfless, S. J., Dao, M., Niles, J. C., Suresh, S., and Han, J. A microfabricated deformability-based flow cytometer with application to malaria. *Lab on a Chip*. 2011; 11(6):1065. DOI: 10.1039/c0lc00472c. 59

[187]   Rosenbluth, M. J., Lam, W. A., and Fletcher, D. A. Analyzing cell mechanics in hematologic diseases with microfluidic biophysical flow cytometry. *Lab on a Chip*. 2008; 8(7):1062–1070. DOI: 10.1039/b802931h. 59

[188]   Solmaz, M. E., Sankhagowit, S., Biswas, R., Mejia, C. A., Povinelli, M. L., and Malmstadt, N. Optical stretching as a tool to investigate the mechanical properties of lipid bilayers. *RSC Advances*. 2013;3(37):16632–16638. DOI: 10.1039/c3ra42510j. 59, 60

[189]   Guck, J., Ananthakrishnan, R., Mahmood, H., Moon, T. J., Cunningham, C. C., and Käs, J. The optical stretcher: A novel laser tool to micromanipulate cells. *Biophysical Journal*. 2001; 81(2):767–784. DOI: 10.1016/S0006-3495(01)75740-2. 60

[190]   Guck, J., Schinkinger, S., Lincoln, B., Wottawah, F., Ebert, S., Romeyke, M., Lenz, D., Erickson, H. M., Ananthakrishnan, R., Mitchell, D., Käs, J., Ulvick, S., and Bilby, C. Optical deformability as an inherent cell marker for testing malignant transformation and metastatic competence. *Biophysical Journal*. 2005; 88(5):3689–3698. DOI: 10.1529/biophysj.104.045476. 60

[191]   Bellini, N., Bragheri, F., Cristiani, I., Guck, J., Osellame, R., and Whyte, G. Validation and perspectives of a femtosecond laser fabricated monolithic optical stretcher. *Biomedical Optics Express*. 2012; 3(10):2658. DOI: 10.1364/BOE.3.002658. 60

[192]   Ekpenyong, A. E., Toepfner, N., Fiddler, C., Herbig, M., Li, W., Cojoc, G., Summers, C., Guck, J., and Chilvers, E. R. Mechanical deformation induces depolarization of neutrophils. *Science Advances*. 2017; 3(6):e1602536. DOI: 10.1126/sciadv.1602536. 60

[193]   Marszalek, P. and Tsong, T. Y. Cell fission and formation of mini cell bodies by high frequency alternating electric field. *Biophysical Journal*. 1995; 68(4):1218–1221. DOI: 10.1016/S0006-3495(95)80338-3. 60

[194]   Ashkin, A. Acceleration and trapping of particles by radiation pressure. *Physical Review Letters*. 1970; 24(4):156–159. DOI: 10.1103/PhysRevLett.24.156. 61

[195]   Guck, J., Ananthakrishnan, R., Moon, T. J., Cunningham, C. C., and Käs, J. Optical deformability of soft biological dielectrics. *Physical Review Letters*. 2000; 84(23):5451–5454. DOI: 10.1103/PhysRevLett.84.5451. 61, 62

[196]   Barton, J. P., Alexander, D. R., and Schaub, S. A. Theoretical determination of net radiation force and torque for a spherical particle illuminated by a focused laser beam. *Journal of Applied Physics*. 1989; 66(10):4594–4602. DOI: 10.1063/1.343813. 61

[197]   Ashkin, A. Forces of a single-beam gradient laser trap on a dielectric sphere in the ray optics regime. *Biophysical Journal*. 1992; 61(2):569–582. DOI: 10.1016/S0006-3495(92)81860-X. 61

[198]   Ashkin, A. Optical trapping and manipulation of neutral particles using lasers. *Proceedings of the National Academy of Sciences*. 1997; 94(10):4853–4860. DOI: 10.1073/pnas.94.10.4853. 61

[199]   Ferrara, L., Baldini, E., Minzioni, P., Bragheri, F., Liberale, C., Fabrizio, E. D,. and Cristiani, I. Experimental study of the optical forces exerted by a Gaussian beam

within the Rayleigh range. *Journal of Optics*. 2011; 13(7):075712. DOI: 10.1088/2040-8978/13/7/075712. 62

[200]   Wottawah, F., Schinkinger, S., Lincoln, B., Ebert, S., Müller, K., Sauer, F., Travis, K., and Guck, J. Characterizing single suspended cells by optorheology. *Acta Biomaterialia*. 2005; 1(3):263–271. DOI: 10.1016/j.actbio.2005.02.010. 63

Authors' Biographies

Wenhui Wang received a B.E. in 1998 and an M.E. in 2001 from Beijing Institute of Technology, and Ph.D. in Mechanical Engineering from National University of Singapore (2005). He received post-doc training at the University of Toronto and then joined the faculty at the University of Canterbury in 2007. In 2012, he relocated to Tsinghua under the Chinese Government Young 1000-Talent Plan. His current research interests include BioMEMS and microfluidic devices and systems, aiming for bio-micro-manipulation and analysis of single cells and model organisms.

Liang Huang received a B.E. in 2008 from Tianjin University, an M.E. in 2011 from University of Science and Technology of China, and a Ph.D, in 2019, from Tsinghua University. H is now at the Hefei University of Technology, where his current research interests include BioMEMS and microfluidic devices.

Printed in the United States
by Baker & Taylor Publisher Services